Global Emergency Actions

For a Small Urban Industrial Planet

Alan Wittbecker

Books by Alan Wittbecker
- *Eutopias: A Poetic Commonwealth of Earth* (OP)
- *Ordering Spaces and Living Places: Aesthetic and Ecological Dimensions of Place* (OP)
- *Poetic Archaeology of the Flesh: Creative Language, Physics and the Ecology of Being*
- *One Earth Many Worlds: The Role of Cosmologies on Ecological Impact and Accommodation*
- *REviewing REthinking REturning: Essays of Life, Ecology, and Design*
- *Good Forestry from Good Theories and Good Practices: Essays on Ecological Forestry and Ecological Design*
- *Eutopian Essays: Towards Making Good Places with Thought Experiments & Ecological Designs*
- *Eutopias: Making Good Places Ecologically & Culturally Using Thought Experiments & Informed Designs*
- *Domiture: The Coevolution of Nature & Culture* (OP)
- *Global Government*
- *Radical Ecological Thought Experiments*
- *Redesigning the Planet: Foundations*
- *Redesigning the Planet: Local Systems*
- *Redesigning the Planet: Regions*
- *Redesigning the Planet: Global Ecological Design*

Global Emergency Actions

For a Small Urban Industrial Planet

In a eutopian framework based on
forms & values of traditional cultures
and on applied ecological systems reasoning
using ecological design & thought experiments

Third Edition

Alan Wittbecker

Urania Science Press
Sarasota
2015

Published by Urania Science Press. Mozart & Reason Wolfe Ltd.
   SynGeo ArchiGraph
   8051 North Tamiami Trail, No. 30
   Post Office Box 370
   Tallevast, Florida 34270-0370
   editor@reasonwolf.com

For more information on sites and projects in the text:
   SynGeo ArchiGraph Co.: www.syngeo.org
   Ecoforestry Institute: www.ecoforestry.net
   G. P. Marsh Institute: www.gpmi.us
   Pan Ecology: www.panecology.net
   Rian Garcia Calusa: www.riangarciacalusa.com
   Eutopian Ecologists: www.eutopias.net

Copyright © 1970, 1976, 1984, 1990, 1994, 2000, 2007, 2012, 2014, 2015
   Alan Wittbecker

All rights reserved. No part of the book may be reproduced in any form or by any means, including information storage and retrieval systems, without the prior written consent of the Author or Publisher, excepting brief quotes used in reviews.

Publisher's Cataloging in Publication Data
Alan Wittbecker 1946-
   Global Emergency Actions

   Includes Index.
   1. Man—Response to catastrophe. 2. Human Ecology. 3. Ecological Design & Planning.
   I. Title.
   GF75.W5853 2015

ISBN-13: 978-1505677478
ISBN-10: 1505677475

Book Design by Rian Garcia Calusa
Printed in the United States of America
Third Edition January 2015

10 9 8 7 6 5 4 3 2 1

Dedications

To Garrett Hardin, for his support and correspondence, as well as for his criticism, commiseration and Darwinian postcards. I have followed his lead on many things, from tackling unpopular topics to supporting the Hemlock Society, and I am grateful that he made the path easy to follow.

Special thanks to John B. Cobb Jr., for his constant support and encouragement, and for inviting me to speak at the Center for Process Studies at Claremont (1996). All I really wanted was for him to write this book; I know that he would have done a better job.

Special thanks to Michael W. Fox, for showing me that being a pariah can be an honorable thing, and then for blazing that path. I have been honored to work on some of his veterinary projects. He offered me more help and inspiration than I can ever repay.

Thanks to Arne Naess for hiring me to work on the Wolf Project and to lecture at the Center for Nature at the University of Oslo one spring (1987), between boxing, hiking, climbing, and swimming expeditions, and while discussing the relevance of ecology and philosophy to peace, prosperity, equity, and balance.

Thanks to Norman Bowie for inviting me to work on this book at the Center for the Study of Values at the University of Delaware one summer (1986) as a Visiting Scholar, and for including me in meetings and letting me continue discussions with faculty (from 1972).

Thanks to Paolo Soleri for giving me the opportunity to conduct arcological thought experiments for two summers (1983, 1992 and a week in 2002) at Arcosanti and for including me in design discussions. I still believe that is the best place on earth to live.

Thanks to Eugene Odum for his many comments after our conference presentations in 1983, 1986 and 1998. As a graduate student in ecology I learned from his textbook. As an instructor and ecosystem ecologist, I taught my first courses using his textbook.

Thanks to Alan Drengson for publishing and promoting my articles, as well as for arranging for many of his students to take my first class in ecological forestry (1994), and for contributing his work to my books and journal (*Pan Ecology*). I am grateful for his help, friendship and correspondence.

And, of course, my continued gratitude to Michael Barnes, Neil Evernden, Buckminster Fuller, Twila Jacobsen, Neil Keefe, David Klein, Nadya Kristoforova, Devorah Bell, Boyd Martin, Linda Schapf, William Odum, Nela Rachevitz, Paul Shepard, Henryk Skolimowski, Emerson Wittbecker, Margaret Wittbecker, and Precious Woulfe for their unselfish criticism, suggestions, support, or assistance. Finally, if I had not been trapped in a mountain snowstorm at the observatory on Mt. Lemon in early 1973, I might not had time to read books by Theodore Roszak, Ivan Illich, and Leopold Kohr, and become determined to continue writing about these topics.

Contents

Preface: Emergencies Now   7

Nature of Catastrophes: A Brief Introduction   11
  Platopias: Flatscapes from Failures & Losses   11
  Utopia: No Place   14
  Topopoetics: Making Places   16
  Eutopias: Making Good Places Ecologically   20
    & Culturally by Design & Practice
    Responsibilities of Individuals   22
    Responsibilities of Nations   23
    Responsibilities of a Global Framework   23

Global Emergency Actions   24
  Getting to Good Places
    Adapting a Psychology of Catastrophe   28
    Taking Immediate Action
  Taking Coordinated Action on Three Levels   30
    Actions as an Individual/Community   30
      Individual Actions: Health, Consumption
        Jobs, Participation, Control, Service
      Community Actions: Getting Income: Taxes,
        Licenses, Laws. Giving Outgo: Income &
        Medical Vouchers,
    Actions Contributing to a Nation   60
      Secure borders, Establish rights, Survey,
      Monitor, Plan, Restore, Conserve, Balance
      Budgets with New Taxes, Create Jobs
    Actions Contributing to the International   124
      Global government, Owned Commons,
      Complete disarmament, Make budgets,
      Identify goals & threats, New agencies

Why it Could Work: A Brief Conclusion   167
  The Dance of Art Money & Ethics
  Creating & Maintaining Eutopias Now   171
  What Eutopias Can Do   173
    Eliminate Bad Approaches & Actions, Integrate
    Tools & Designs, Increase Understanding, Start
    Making Good Places, Start a Revolution
  Moving Forward (& Backward, Inward & Outward)   185

Ecodex (Summary of Eutopian Code & Actions)   188
Appendix   194
Author   195

Global Emergency Actions
For a Small Urban Industrial Planet

Preface: Emergencies Now

What emergencies? These emergencies are pressing and concurrent: Ecosystem conversion, simplification, collapse, and destruction by agriculture; the conversion of agricultural areas into cities and roads; animal and plant interferences and extinctions; the creation of large quantities of waste; the disruption of elemental and biogeochemical cycles; the misuse and overuse of resources, especially minerals and fossil fuels; the narrowing of agricultural diversity and species; the overuse of energy, pesticides and fertilizers; the continued use of exotic and toxic materials and chemicals; the fragmentation of wild systems by harvesting and by roads; the narrowing of cultural capital and languages and the extinction of many; the economic betrayal of trust; the global homogenization of products and lifestyles, and the extinction of styles and alternatives; the continuation of slavery, violence and war; gross inequities in wealth between people and nations; the increased use of energy and reliance on fossil carbon; and, the refusal to consider climate change or respond to it. Unaddressed, these emergencies are converging into one gigantic emergency that is going to create one gigantic global catastrophe.

    In the case of industrial nations, we have been embracing excess for many generations, so that we are crippled by stress and sickness. Perhaps all we need is a diet. The essence of a diet is to restore one's self to health, by restricting unhealthy consumption. As societies and cultures may also be guilty of this kind of behavior, so they need to put themselves on a diet. Archaic and agricultural nations have been strapped by historical inequities and unfair trading. Their challenge is to avoid simply repeating the same errors and consequences in the rush to acquire minimum standards and wealth. The solutions for all nations include trying new kinds of balance for self-reliance, paying attention to cultural and physical catastrophes, and striving for better equity. Because of the extent of our overuse and ecosystemic conversions, and their effects on natural sources, this situation is an *emergency*.

The nature of an emergency requires everyone to drop their normal activities and normal behaviors and to respond to a catastrophe. The catastrophe is usually quite evident, a wall of fire or a massive surge of water that will destroy or has already destroyed homes and people, as well as insects and birds, plants and animals, and their habitats. But, we are finding that not all catastrophes are fast, human-scale or visible. The effects of those large changes make us uneasy but not adrenaline-ready; the changes are reflected in starving children, hotter summers and stronger storms, failing food supplies, and collapsing infrastructures. We seem reluctant to identify the causes of these catastrophes, the status of real emergencies, partly because the catastrophes seem like natural events, such

This book is a list of actions that should be taken to respond to our current global emergency, which is a result of developmental changes in the planet in its solar environment, as well as in the development of climate and the biosphere, and furthermore with the development of human cultures that have accelerated changes to the composition and stability of ecosystems and human settlements — to avoid catastrophe. Many catastrophes are long-term, large-scale, slow, incomprehensible, and invisible, and they have been developing for thousands of years. This requires a change in our perception and adoption of an attitude of catastrophe, which may be necessary before we can take coordinated actions. It is also unlikely that we can respond just as individuals or turn over responsibility to a single agency or government. We have to work on every level, from individual and community to nations, confederations and a global government.

Most of these suggested actions are based on a general knowledge of ecological change and human development. Many are based on the historical stability of traditional cultures, which have survived within a variety of limits, including weather, vegetative shifts and overuse of resources. Some are based on preliminary studies of long-term environmental changes and on human adaptability. Some are based on global ecological designs. A few are intelligent guesses, and a few may be expressions of a wild hope.

Some of the actions may seem contradictory and counter-intuitive. Others, such as some new forms of taxes or rules of exploitation (especially regarding pollutions and biocides), are only temporary and are meant to achieve a quick effective balance; when they are no longer necessary or useful, they can be removed. And, still others, such as an ecological basis of exploitation and distribution, recognition of real ecological, psychological, and social limits, and new forms of conflict resolution, may be always necessary as part of a complete human ethics, economics and politics.

These actions depend on a massive, cooperative effort. They promise a greater equalization of opportunities to use resources that are becoming more rare and expensive. They use openness and a joint sense of responsibility. However, these are just suggestions to be tried and refined — then discarded or supplemented by others. Any plan for emergencies has to be flexible and adaptable. It is crucial to avoid committing to actions that make things worse.

This effort does not mean that we should stop working or give up our jobs and dig holes to hide in, but it does mean that some jobs and some industries should be abandoned, and others need to be created and applied; and it means that every effort needs to be tied to the long-term goal of designing and making the planet habitable for all forms of life. It does entail sacrifice, but most sacrifice can be borne if everyone is participating and if the goals are shared and desirable. As with any medical approach, we have to let the planet and biosphere do most of the work to be successful, although our participation, restraint and cocreation are critical, now.

We have to learn to recognize and respond to these slow catastrophes, these invisible catastrophes and these very large and long-term catastrophes. And, we have to do it now, before they crest and become overwhelming. We can do it. We have the evidence that things are taking a downward turn (the original meaning of the word catastrophe). We have acted on a large-scale before, in times of a world war. We were able to treat war as an emergency and to encourage or enforce remarkable changes, such as rationing or job-remolding. We were able to take these actions without destroying our citizens or our cultures.

Although nature, and our human nature, are not enemies to be vanquished, the current situation has similarities to war. Massive changes threaten our lifestyles. Resources are removed from our reach by thoughtless or inefficient use. Changing insect and animal populations seem to be attacking our food supplies. Species being forced to extinction. Habitats are collapsing; Dangerous chemical wastes are accumulating. Ozone holes are growing, extreme climates pressing, and the entire planet seems to be wobbling. Changes in climate and ocean balance, as well as renewed diseases and infiltrated toxic chemicals, threaten our lives. And, it is happening everywhere, at once.

We have been fooled by the fact that we cannot see an enemy. We have been misled by the slowness and subtlety of the penetration of our defenses. We have been betrayed by our own desire to continue our industrial dreaming at any cost. Some people have noticed changes and have been crying alarms, but they have not been loud enough or persuasive enough. Everybody needs to be awakened; everybody needs to participate, everybody needs to sacrifice and work towards peaceful solutions.

The big problems seem insurmountable, and simple actions will not save our civilization from catastrophes. Part-time participation will not be enough to reverse the degradation of ecosystems, and partial business greening will not stop the unraveling of global cycles.

What is needed is an immediate, comprehensive approach to this situation—combining our actions in a global framework. The framework, could preserve what is good and useful in human cultures and sciences, and reserve what is necessary for nature to keep regenerating itself, while addressing the cascading problems of the modern expansion and development with an emergency approach. A practical framework could allow the creative anarchy of traditional-size cultures to be able to implement appropriate technology to deal with their resources and with other cultures through a revitalized and empowered international body that has the power of taxing global resources and properties for its own support, as well as the power to disarm and neutralize the unhealthy influences of large nations and corporations. The framework would limit human expansion to domestic and artificial areas, by specifying responsibilities and duties, while permitting the free operation of nature on the majority of the planet. It would save neopoetic areas and reserves wilderness. It would encourage respect for natural and cultural capital. It would recommend recognizing limits and planning for them using an ecological perspective and a metaphorical approach—it is metaphor-based as well as science-based, and limits-based as well as culture-based. This framework would be concerned with saving human cultures and the environments that human cultures have come to fit in comfortably.

Although this framework has not been named, yet, it can be described: It has to be global and local; everyone has to participate; it has to recognize ecological limits; it has to recognize human limits, and it has to recognize and respond to the bad trends. The framework could be utopian or eutopian. It could be democratic or anarchistic. It could be scientific or religious. But, we have to create a first conscious global frame, now, and apply it to our problems and challenges.

Why would such a framework be successful? Because life has over three billion years experience with changing and adapting, because human life and cultures have over 50,0000 years of practical experience adapting and making changes, and because humans are immensely adaptable—if they can adjust to pollution, poverty and suffering, then they can adjust to a few good changes.

Perhaps it is already too late—perhaps limits have been passed and the catastrophes cannot all be reversed. We do not know, and may never know, but we can still act as if we were wise, as if doing the right thing makes a difference. And, we will have worked together to help others, to improve things and to make good places. If we act *now*, this month, this week, this day, *this hour*!

Figure 1. *Map of Proposed Palouse Grassland Reserve*

Nature of Catastrophes: A Brief Discussion

Platopias: Flatscapes from Challenges Failures & Losses

The local environment, as well as the region and planet, offers opportunities and challenges to all living organisms. How humans respond to these things results in accomplishments or problems, in creations or deaths. Every year numbers are collected in every area of human interest. Those numbers indicate the deterioration of water and air quality, the erosion of soil and land, the destruction of forests, the decline in health and longevity, the deaths of people, the fracture of cultures, and the wobble of planetary cycles. In the past years, the numbers have worsened dramatically.

The numbers of changes are incredible: For the first time in human history, in the year 2000, as many people lived in large communities (over 20,000) as in small communities (under 20,000). Sixty percent of Earth's inhabitants are expected to live in cities by 2030, according to the United Nations—the same year global carbon dioxide emissions are expected to increase by almost two-thirds of what they are today. By 2025, worldwide energy consumption is expected to grow by 54 percent, while worldwide oil production is predicted to begin declining in 2016.

The numbers of differences are revealing. The inequity between people continues to increase. In 1996 the UNDP estimated that the wealth of the world's 358 billionaires exceeded the combined income of nations that are home to 45 percent of the world's people. And the gap grows. If we added millionaires, what would that be in terms of power or disgrace?

The numbers of human deaths are far more disturbing. There have been massive human die-offs in the past 106 years since the start of the Twentieth Century. These numbers tell a story of big death. From democides—the intentional killing of races, nations, tribes or communities—total deaths may range from a minimum of 150 million to a possible 350 million people, who were shot, knifed, burned, suffocated, poisoned, starved, crushed, drowned, hanged, bombed, or buried alive, in a plague of violence. This includes Joseph Stalin's 1932-3 forced famine that killed 7 million people, as well as the Nazi holocaust (6 million) of Jews and Gypsies under Hitler by 1945, the current deaths in the Congo (by 2006, 4 million), Pol Pot's Democratic Kampuchea, Khmer Rouge 1975-79 (over 1.5 million), Sudan (1985-2004, 2 million), Ethiopia (1978, 2 million), North Korea (1948-87, perhaps 2 million murdered), and Pakistan (1940-50s, 1.5 million murdered). Rwanda (1994) and Bosnia-Herzegovina (1992-5), add 800,000 and 200,000 to the total. Is it possible to even imagine that number?

Deaths as the result of formal or declared wars are surprising less than democides, which are often internal to a nation, at about 61 million dead. The war between China and Japan, before and during the second world war (1932-1945), resulted in 14 million dead. The rest of the world may have experienced twenty one million dead during the six war years.

Famine, once the greatest producer of deaths in agricultural nations for thousands of years, may have killed 32 million in this century.[1] One of the largest modern famines was the Great Famine in China, centered in the early 1950s; it killed 10.7 million people. There were two famines in India, in 1942 and 1965, that each killed 1.5 million people from drought. Two famines in Korea, in 1948-87 and 1995-8, each killed over 1 million people.

Disease, once the greatest producer of deaths during early phases of globalization, has killed many millions, especially at the end of the Roman empire, when the Justinian plague in 540-590, might have killed 100 million; the Black Death of the 1300s in Europe and Asia, and the beginning of the Spanish Empire, produced 46 million dead. The influenza pandemic in 1918 killed over 20 million people, and possibly as many as 50 million. The recent AIDS epidemic, from 1978 to 2001, counts for over 23 million.

Disasters, from drought, flooding, earthquakes and other regular planetary events, killed 21 million people. Accidents, from transportation from horses to space shuttles, killed half a million people. Murders and terrorism killed hundreds of thousands (perhaps 0.02 million). Nations, such as the U.S. or the United Kingdom, have killed thousands of innocents; the U.S. is guilty of indiscriminate bombings of Germany and Japan in 1945, then later in parts of Africa and the Middle East, and the U.K. caused thousands of deaths with its 1914-1919 food blockades of Germany and the Middle East. Such a list would also have to include Afghanistan, Angola, Albania, Ethiopia, Burundi, China (1917-1949), Croatia, Czechoslovakia, Indonesia, Iraq, Turkey (1919-1923), and Uganda.

Many of these overlap by category. For instance, ethnic cleansing can start from changes in the distribution of food, that lead to famine and disease. Poverty is never listed as a cause of death, but over 15 million, perhaps as many as 30 million people, die from a lack of clean water, food, medical services, or shelter, every year, including 2006.

The numbers on the living environment are disturbing and critical: The list of animal deaths in the United States[2] in 1984 reads like a doomsday book of atrocities: 22,078 North Pacific fur seals clubbed to death; 17 million mammals trapped for fur in the U. S. — 303 million throughout the world; 12 million unwanted pets put to death; 70 million laboratory animals used in experiments; 3.5 billion chickens killed for food; 700,000 cattle dead from transport-related injuries; 598,757 animals shot for sport on wildlife refuges. Although species are still being identified at the rate of 8,500 new insects species and 100 new fish species per year,[3] probably 400 species are driven to a premature extinction and 1,000,000 species are threatened every year.[4] Statistics for habitats are almost incomprehensible. Three billion cubic meters of wood are consumed annually. Twelve million hectares of forest are cleared annually and 10 million hectares are degraded. Marsh lands are filled in; coral reefs are mined; and grasslands are paved over.

Possibly 59 percent of arable land degraded; 50 percent of fresh water co-opted for human use; 50 percent of the planet's wetlands modified, drained or destroyed; 50 percent of the coral reefs damaged and perhaps 20 percent destroyed; extinctions that are uncounted, perhaps uncountable. We tend to think of problems highlighted by numbers as unwanted 'side-effects' of the wanted main-effects, but all effects are equal, as Buckminster

Fuller noted, and must be addressed as equal. A problem—from the Greek words 'to throw forward,' which is what we tend to do with them—can be considered as a question proposed for solution. Most things identified as problems are embedded in a network. Nothing is simple; there is not one problem, there is not one solution. Problems could be considered also as challenges that we must respond to continuously, in the process of living, not as puzzles that have to be solved once for all time. A challenge is a calling into question or a demanding task—a challenge is defined as 'a call to take part in.' It is about consciously choosing to see what can be done, rather than dismissing a conflict as terrible and unsolvable. When challenged by some situation, we react by habit, although this may be disconnected from other habits. Habits protect us from many problems. Addressing a problem often has to do with a power struggle, which becomes part of the problem. If problems are regarded as challenges that require a social response, then much of the conflict can be avoided.

The problems of cultures, and of modern, industrial, corporate, urban civilization, have been documented quite thoroughly. We have identified most of the problems in the 'problematique,' from erosion, pests, and fertility loss, to population migration and diseases, and we have addressed them separately and partially, using technological innovations or political adjustments. But, we have not dealt with them in a whole pattern. We have not understood them as complex large dynamic systems.

Sometimes we forget, however, that the decisions of our ancestors can saddle us with losses, just as our losses will encumber our heirs with deforested landscapes on depleted soils, despoiled by exotic chemicals and hazardous wastes, in a network of impoverished habitats with an unstable climate, and of course, compounded by large intergenerational financial debts. This network of problems can be grouped under eight large categories, each of which contains a multitude of related problems.

These problems are responsible for eight significant and critical losses. First, there is the loss of nature and the wild, the loss of the whole and the pieces, ecosystems, habitat, species, and individuals. Second, the loss of culture leads to the inability to adapt to unique environments due to unstable social environments. Third, the loss of health results in weakness and sickness, the inability to maintain the self or to produce necessities. Fourth, the loss of fitness leads to the inability to function under normal environmental conditions and to reproduce. Fifth, the loss of equity results from the massive breakdown of the distribution of goods and luxuries. Sixth, the loss of renewal reflects the inability of social systems to renew themselves or provide security and resources for their constituents. Seventh, the loss of accord happens when people or cultures are unable to work together or control conflict. And, eighth the loss of design is the inability to imagine, shape and build things that enhance life and safety; it is the inability to respond to changing circumstances. Each category extends numerous threads to the other categories.

Utopia: No Place

Utopias as visions of ideal societies can be found in almost every culture: In prophecies, visions, dreams, myths, and ideologies. George Orwell suggests that the utopian vision has been consistent since Plato, with its dream of justice through reason, through the elevation of public life, and through the regulation of private life, community property, and good breeding. Utopias are not just visions of noplace, but of otherness. The ideas of a golden age and an ideal city were combined in modern western utopias. The discovery of new continents opened new possibilities for utopias. We became enchanted with size and speed, and tried to create ideal places.

Grand Thefts
Despite our ideals, bigness and speed, most of the world's population goes hungry; even fewer are fulfilled as actualized human beings. The utilitarian aim of greatest good for greatest number has been vulgarized to mean the greatest number of goods for those who can afford them. In our attempts to manufacture the good life for those who can afford to buy it, we have deprived everybody of clean air and water, quiet nights, darkness, open spaces, and other indefinable qualities. Soils are destroyed, wildlife is killed. We devour nature to assuage our disease; we try to fill our emptiness with goods. We can only gain past a certain point before our gain causes the loss of something else that we need to be healthy.

Modern technological society ravishes nature and mutilates humanity with the products of its materialism. Industrialization has distorted people's lives and cheated them of bread and justice. Oppression darkens the mind and narrows the spirit. In a mass consumption society, people impoverish themselves spiritually while impoverishing others materially. This is theft. As with the Christian ten commandments, for example, most loss can be reduced to theft, whether of a life, mate or name. Most of our modern problems can be considered the consequences of forms of theft, such as of life, common sense, comfort, security, and finally choice.

Achieving Placelessness
Due to this combination of thefts and losses, magnified by that combination, we have created the opposite of peace, and the opposite of place. A. De Tocqueville had previously identified a trend of human uniformity in the 1830s, when he wrote that variety was disappearing, and the same ways of acting, feeling and thinking were copied around the world. A number of scholars, including Christian Norberg-Schulz and Edward Relph, have noticed that we are creating flatscapes, devoid of depth and providing only one mediocre possibility, a chrome-plated chaos.

Placelessness begins with adoption of an attitude, an abstract, geometric view. With an inauthentic technique, places can be treated as interchangeable and unremarkable, where nothing is significant. Cutting historical roots and eroding symbols contribute to an awful placelessness, an alienation to place, an inability, finally, to have a home and to live there. This becomes the fate of millions, and it increases.

After shaping ourselves to technology and necessity, we have lost the knowledge of how and for whom to care. We have learned not to care, to be dispassionate (uncaring), unattached (placeless), and objective (uninvolved). Other animals have used languages and tools, so it is not those things alone that account for the lost knowledge. We 'not-care' because we are confused. Our confusion results from being out of place and not having an identity.

People use to identify with place. The gain of a global monoplace has led to a loss of identity. The mass production of homes, as well as of music and art, has led to fewer creations. The commonness of culture, at a low common denominator, has led to fewer identities to choose from.

The Scope of Failures

Acceptance of limits is not a kind of failure. Awareness of inadequacy, as of ignorance, is a positive accomplishment. True failure is indifference to inadequacy. The failures in our personal, community or national character can be seen to be responsible for the problems identified by Konrad Lorenz as the seven deadly sins of civilized humanity, from destruction of nature to the loss of civility. These failures can be described as a series. The failure of perception is the collapse of our ability to see, understand and know the large and less visible patterns of nature. The failure of intelligence is characterized by the reluctance to use knowledge in new situations, the reliance on belief and myth not analyzed by common sense or logic. The failure of imagination is the indifference to whole images and good ideas and their connections with living contexts. The failure of integrity flows from the willingness to be corrupted by selfishness and desire, allowing one's identity to remain small and disintegrated. The failure of will results from the refusal to make reasoned decisions or to act with purpose if the need to do so is inconvenient or bears any cost; it also means being paralyzed with fear. Finally, the failure of charity betrays a disinterest in sharing or giving, detachment from the needs of the social and wild communities, and a lack of generosity.

Of course these failures occur in endless combinations. Maybe the failure of our modern civilization is a failure of imagination compounded with a failure of nerve. We cannot imagine an alternative to war, and we cannot act beyond emotion. We cannot imagine beauty in the old, messy nature, and we are afraid to try to do without luxuries or to try to sacrifice anything to change the momentum of industrial civilization. Our false models and ideas, combined with our failures of imagination and charity, for instance, may doom us to make only flatscapes and noplaces.

Are we doomed by our failures to live in nowheres and noplaces, to have the same dull jobs and the same minor rewards? Are we doomed by our failures to continue to destroy the habitats and ecosystems that support and nourish us?

Topopoetics: Making Places

Or can we change? We make the places we live in, or rather, we remake places by adjusting them to fit us. We also fit into a place and make images of a place that also fit, that is, we adjust ourselves as well. We are a place-making species; human beings are topopoetic, like most species. We have to stand in place before making or remaking a place. As human scale changes, however, our species acts more like an agent of corruption. And, as we become more successful, in the senses of reproduction and domination, we create conditions that function like traps, that limit our options.

Humanity as an Agent of Change
Regardless of our failures and the losses they cause, and because of our large population size and energy use, and the effects of our domestic plants and animals, we have become an agent of change, as much as any force of nature. Certainly geological processes, as well as solar system effects, such as the output of the sun or meteorites, are forces of nature. Certainly, the environments, including climate and oceans currents, are forces. Human actions, such as deforestation, identified by John Perlin as perhaps the main reason for the decline of the Hittites and Babylonians, have the effects of forces. Desertification, according to Uwe George and others, is another. Disease patterns, according to W. H. McNeill, are crucial. To disease, or germs, Jared Diamond adds steel and guns as forces that have shaped human history and societies.

By their activities, human beings change the places they live. Much of the change is easily incorporated in the cycles of renewability of the ecosystems. However, humans often change the directions of such systems by simplifying or degrading the systems. In this case humans act as agents of interference.

One analogy of humans as such agents is as a parasite: A consumer feeding on another living organism, usually inside, drawing nourishment and weakening the host. States acted like macroparasites, according to William McNeill, but becoming less violent or unpredictable over time, as they adjusted to their host populations.

As crowding increases competition, there is more pressure on remaining wild areas. The human system parasitizes humanity and nature. Humanity becomes an autoparasite, a new pseudo-species. Technology enlarges the number of niches for us; tools fit humans to different habitats, displacing other species. We steal from animals and plants, from the earth, and from our own descendants. Hobbes foresaw this kind of war of each against all. The systematic destruction of human beings and animals is not an isolated peculiarity. A fat parasite often kills it host and then dies itself. Perhaps, humanity is an agent of a different sort, a systems agent that encourages only positive feedback.

Perhaps human expansion is like a cancer, as Alan Gregg suggested, when he compared the world to a living organism and the explosion in human numbers to the proliferation of cancer cells. He sketched other parallels between cancer in humans and humans' cancer-like impact on the world. Cancer cells proliferate rapidly and uncontrollably in the body; humans continue to proliferate rapidly and uncontrollably in the world. Crowded cancer cells harden into tumors; humans crowd into hardened cities. Cancer cells infiltrate and destroy adjacent normal tissues; urban sprawl devours normal open land. Malignant tumors shed cells that migrate to distant parts of the body and set up secondary tumors; humans have colonized just about every habitable part of the globe. Cancer cells lose their natural appearance and distinctive functions; humans homogenize diverse natural ecosystems into artificial monocultures. Malignant tumors excrete enzymes and other chemicals that adversely affect remote parts of the body; humans' motor vehicles, power plants, factories and farms emit toxins that pollute environments far from the point of origin.

It is not in a self-interest of a tumor to steal nutrients to the point where the host starves to death, for this kills the tumor as well. Yet, tumors commonly continue growing while the victim wastes away. A malignant tumor usually goes undetected until the number of cells in it has doubled at least thirty times from a single cell. The number of humans on Earth has already doubled thirty two times, reaching that mark in 1978 when world population passed 4.3 billion. It is over seven billion now, on the way to the thirty third doubling to 8.6 billion. After thirty-seven to forty doublings, at which point a tumor weighs about one kilogram, the condition is usually fatal. The question is: Has our doubling been fatal, without our knowing it — large complex systems may take a long time to collapse — or does the system have more flexibility than an organic body?

The metaphor of cancer may be more appropriate than a shadow, ghost acreage or ecological footprint. After all, a live footprint can stimulate some kinds of ecosystems, such as shortgrass prairie. What humanity does is transform the ground under the footprint into a new, poorer system.

Perhaps humanity is simply a dominant ecological agent. A dominant is a species with greater influence than any other in its biotic community, changing the lives of other species and the character of the habitat. Humanity is a pandominant species. As such, humanity reclaims, overgrazes, clears, depletes, and wastes at a level that threatens the stability and existence of many systems. One of the ecological consequences of human activity is the degradation of wild habitats for human developments and the introduction of novel elements into the biosphere — elements that have not been harmoniously worked in over time. The biomass, or demomass, of the human species probably far exceeds the biomass of any nondomestic species, and that biomass is supplemented by the tremendous biomass of domestic animals, which is four times greater.[5] This biomass forms an equivalent population that consumes much of the same food, such as milk, fish, and grain.

This pandominance has major effects on ecosystems: Transient perturbations in energy relations, from oil spills or burning, for instance; chronic changes and shifts of systems, from dams, irrigation, or chemical wastes; species manipulation, from the import and export of exotics; and, interference competition with wild species, as opposed to exploitative competition, which can be stabilizing. No single change is exclusive to humans as a species, but they are excessive, rapid, compounded, and large-scale. There is movement of soil, but also massive erosion. There is movement of minerals, but also disruption of mineral cycles. There is the addition of novel elements into the atmosphere, but there is also a massive release of carbon.

Our pandominance results in our becoming a geological and climatic force. When people use more of the earth's supplies in a certain period than can be replenished in the same period by the sun, they are eating into the natural capital. This can create a trap.

The Trap of Success
Cycles that do not operate with the right kind of feedback function as traps. Thus, phosphorus becomes trapped in an ocean sink, and can only be recycled by long geological processes or by specific harvests through human activity. Sometimes, the rearrangement of systems by human groups leads to a position where further rearrangement is not possible. In many cases it is hard to tell if the destructive use of land preceded ecological problems or followed from efforts to maintain production after an ecological challenge. Cause and effect are hard to separate. The same environment that challenges a culture with some kind of change, also offers opportunities with the change. New resources can stimulate economic activity and increase the level of living.

The use of resources by a people, where the replenishment rate is constant and the rate of use exceeds it, is a serial trap. This trap results in ecosystem degradation that is less reversible. The industrial age mistakes the rate of discovery for the rate of recovery.

Agriculture is an energy trap, because it allows a higher concentration of energy, that is, higher yields, but then it requires more energy be put into the system to maintain it. The system has to produce more energy than it uses to be useful and sustainable, with a surplus for trade.

Sedentism is a trap, as people make permanent homes near domestic fields. As the population of sedentary communities increased, the wildlife numbers decreased. The productivity and narrowness of food increased. Thus, there was less possibility of returning to the foraging lifestyle. People became committed to the new lifestyle. Intensity was no longer an option either; it had to be pursued. One problem of sedentism is that the individual cannot simply move away to avoid conflict. People are tied to a particular place and have to communicate to adjust to sharing places.

The city is a different kind of trap, that offers intensity and opportunity, but requires massive imports of supplies to survive. The size and scale of cities create the dual attraction and despair.

Global capitalism can lead to a consumption trap. Capitalism claims to serve the wants of the people, but it spends half its efforts creating more wants in people. Not many of those wants are real, or as real as cereal and roofs. Few of the soft services satisfy real psychological needs.

Markets advance individual desires and not social goals, for instance, by offering running shoes, not inner city restoration. Instead of being free from economic want to develop their potential as creative human beings, people are trapped in a consumer cycle. Self-actualization is postponed for self-gratification. Furthermore, capitalism can undermine traditional cultures by offering consumerism in the place of established, effective guides for behavior. Social roles seem irrelevant by comparison, if the good life can be bought without effort.

Being in a trap means much-reduced flexibility and fewer choices. That is, being in a trap makes one vulnerable to many other changes that could be avoided if one were not in a trap. When the weather got colder, hunters and gatherers could move south. Cities could not. Civilizations are more vulnerable to smaller climatic changes.

Addictions, such as those to luxury foods, or oil or money, make it difficult to escape from a trap, a trap being a kind of energy well or gravity well. Addictions can amplify some emotions, such as fear or hate, especially as they relate to the possible end of the addiction or the threat of that end. Addictions can justify illegal behavior, especially those that seem necessary to continue the addiction. Of course, many cultures are addicted to the illusion of control and power. The U.S. is trapped in the belief that only it, among nations, can bring prosperity and peace to other nations, with trade or violence. Eventually the trap is escaped, or more likely, collapses with its victims.

One possibility of escape is to imagine different approaches and try to design them with ecological knowledge. Images from literary utopias (or 'outopias') can inspire new thoughts about making places that are more open than traps. Images from established good cultural places, Eutopias, can describe responsibilities and pathways that lead to a more balanced exploitation of nature, within its limits.

Figure 2. Drawing of a Proposed Palouse Underground Arcology (Redesigning the Planet)

Eutopias: Making Good Places Ecologically & Culturally

The second meaning of utopia, 'eutopia,' is not used often. It means simply 'good place.' Good places do exist. They can be described, and even consciously created. Some of the traits that make them good can be understood and repeated. A formal compilation of general characteristics of good places, a eutopias, extends the application of utopian thought. Perhaps the number of good places can be increased with understanding of traditional ways and with more effective metaphors. Good places, and good societies, can be partly understood through certain paths of ecology, economics and politics.

We have often moved to places that we have only heard about. If we do not live in healthy places, we want to move to healthy places. Sometimes, we remake our places until they have the characteristics of good places. Good places can be made. People in optimum size groups, with strong ethics and common beliefs and understandings, either by respecting limits or keeping well below them, can make good places through their actions.

Good places are a confluence of good human societies in healthy ecosystems. We can describe some of the qualities of healthy ecosystems. And, we can describe some of the qualities of good human societies. To have good societies in good places will require good images and good actions.

Characteristics of Good Places in a Eutopian Framework
From political character studies to technological promises, utopias have kept close to the contemporary forms of society. The possibilities described for the future seem to be circumscribed by the limits of human imagination. The entire literature of utopia, imaginative as it is, cannot match the actual diversity of cultures for richness or the depth of nature for wonder.

Many archaic societies employ a set of principles, different from industrial cultures, that may be more adaptive to place. Instead of regarding the 'universe as mechanical, humanity as master, and all persons as equal,' in the industrial view, the Yaruru consider the 'universe static and internal, humans sensible to other's wants, and all beings equal;' by contrast, the Navajo consider the 'universe personal and orderly, events primary, and the family first.' The ways that people live in place reflect their principles. For instance, the Yaruru are much less likely to overwhelm their home place with pollution than any industrial culture.

Other modern metaphors can promise more adaptive behavior for industrial culture. A machine metaphor used by Kenneth Boulding, 'the earth is a spaceship' suggests the limits of the earth and the value of its life-support system, but it masks other realities. The metaphor of the spaceship is a closed system model, which leads to inadequate understanding of open, natural systems. The earth is an open system that sustains life. The earth has no single captain with authority. In fact, the image of a spaceship does not fit a large, organic, nonmechanical system. Alas, the earth is not a spaceship. It is far more dangerous and uncontrollable. It is clear that many human behaviors and many human institutions in the past, which may have been appropriate to a large planet, are entirely inappropriate to a small closed spaceship. "We cannot have cowboys and Indians, for instance, in a spaceship, or even a cowboy ethic," Boulding says, "We cannot afford unrestrained conflict, and we almost certainly cannot afford national sovereignty in an unrestricted sense."

Another metaphor in popular use, such as "the earth is a garden," is a better model for reintegrating humanity into a balance with nature, because the garden is a small balanced system directed by humanity, and part of the larger environment, and dependent on it. The rule of the garden is empirical and based on observation: If you do something, then something else happens. Even so, the metaphor of the garden has important limits. Humanity does not have adequate knowledge to direct all of the processes of nature.

In naming a new science of ecology, Ernst Haeckel combined two Greek words (eco-logos) meaning 'the study of the house.' Ecology relates to dwelling, to the frame that contains us. The desire to refine a focus on our problems has allowed the frame of reference to be neglected. This metaphor has turned attention to the whole. But, it too is limited. The house herein is not a construct any more than a spaceship. And, there is not just one house; there are many unique ones with individual characteristics and connections.

Eutopias, as a general description, uses a root metaphor of many places, in different bioregions. This Eutopias is a framework for human cultures, to preserve the unique image that a society needs to guide it and to make it different from others. To be effective, in contrast with the ideal characteristics of ideal cities, a eutopian framework embodies attributes that are compatible to the values and norms of living cultures.

By being attentive to the characteristics of place, of a healthy ecosystem, and those of a good society, a eutopian framework can be described by its own characteristics from groundedness to comprehensiveness. These characteristics are quite different from the characteristics that can be observed with utopias or industrial designs. The characteristics must be preserved by action. Characteristics can be described, but some of the pieces could be missing. The structure is necessary to contain and fit the characteristics, regardless of what is missing.

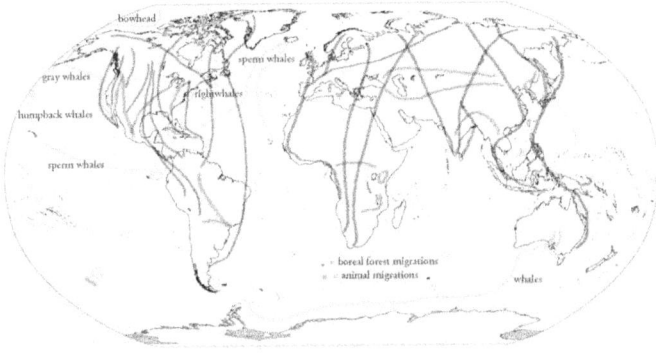

Figure 3. Some Bird & Whale Migration routes (to be saved)

Making Good Places by Design & Practice

People exist in place and act in place. There are many potential levels of action, many ways of being, starting with the individual and continuing through the international. However, individuals are part of families, which are part of groups that often include extended families. Larger groups may include clubs or corporations, communities, states or regions. For the sake of simplicity, communities are considered under the heading of individuals. States and regions are considered under the heading of nations.

Being Individuals in Good Places
From an ecological perspective, living organisms inter-penetrate deeply into nonliving forms and the earth. Individual organisms are woven into a complex fabric. Their activities reshape the fabric. Human beings, to make good places, have to consider their individual and social actions, their participation, as responsibilities.

To make a good place, individual participation is necessary. Participation enters the constitution of place; it is not a fusion where things lose their identity, but a mutual infolding together where each becomes part of the identity of the other. Without it things would have no existence. To exist is to participate in place. Participations are felt, not thought. Participation leads to knowledge and experience.

Many aspects of human social existence can vitiate our responsibility to participate. Aldous Huxley noted that there are three things we center our existence around: Technological gadgets, political power and morality. Each center can be used as a means to solve problems; sometimes people believe that their problems would all be solved if everyone had a radio or computer, were a communist, or were a Mormon. As the collection of gadgets for some grows, the cost to others creates more problems. With the emphasis of political power, only the mechanism changes — from a dishwasher to the corporation or from the toaster to a political party — but, the problems increase. Moral systems allow automatic decisions also, but to no better effect. The idolatries of gadgetry, celebrity and morality occupy our attention. People struggle and suffer in the resulting turmoil. While these concerns are necessary, it would be better to focus consciousness on the unity of the situation, not on single filters of experience.

A rotation of civic, vocational and professional responsibilities would awaken different senses in an individual and round out self-development. Complete society might create complete humans. A eutopian framework could deepen the sense of personal worth, allow spiritual growth, and augment people's need to participate in governing institutions. People are capable of being the most delicate gauge of the health of a place.

Responsibilities of Individuals
Each individual has responsibilities that cannot be evaded or given away. Some of these responsibilities are: To cultivate the self, that is, to be healthy and fulfilled; to engage in meaningful work; and to practice simplicity.

Responsibilities of Communities
The planet is experienced on a smaller frame of reference than global unity or nations; people live on the local level. Local knowledge is knowledge in place, earned in place by generations of inhabitants, through visions and trials, experience, and stories. Thus, individuals are preserved in societies that are preserved in places that are preserved by individuals and societies. Laws, politics, architecture, sports are things of place. They are shaped with local knowledge. A local area is limited by vision, a horizon.

The responsibilities of a community are: To educate people, in schools, corporations, libraries, and museums; to protect people, with public health programs, sanitation, hospitals, and fire departments; to keep the community heritage, through cultural events, shared customs, festivals, art displays and museums; and, to create community wealth, in the form of parks, wilderness areas, monuments, and public buildings.

Education is especially important for women, who are often denied any instruction in language, culture or basic biology. When women are educated, the community becomes wealthier as they start businesses or work in professional positions, such as doctors; birthrates decline as women marry later, sometimes at 32 instead of at 8 years old, and have fewer, healthier children.

Responsibilities of Nations
As protectors of place, Nations have explicit responsibilities, from conserving ecosystems to managing resources and the distribution of rewards and power. Also importantly: To maintain the health of people and culture; to protect its borders, and to coordinate relations, such as trade, with other nations.

Responsibilities of a Global Framework
There is already a world system. A global order is necessary to govern this system. The UN could be dramatically strengthened, or a new body, for example, a Global Union of Commonwealths (GU) could be created. The Global Union should be an elected body and should have the regulatory powers necessary to maintain an healthy global environment. It should have regulatory and advisory powers to maintain the independence and integrity of its constituent nations. It should have regulatory and punitive powers to rectify resource and human rights infringements; only this body would have police powers and impersonal weapons. Various advisory bodies would recommend policies and actions to nations. The Global Union has six basic responsibilities: To ensure a diverse biosphere; to manage resources; to protect unique cultures; to coordinate representation; to provide services to nations; and, to create peaceful conditions. For instance, to ensure a Diverse Biosphere, the GU has to identify, zone, conserve and preserve landscapes.

Emergency Actions

Getting to Good Places

Having regarded the whole of human history, it is possible to see long-term trends or problems. Having discussed what the human relationship to the earth actually should be, and imagined goals and responsibilities, it is time to outline the physical steps that could be used to create a eutopian situation. There is always some path to a destination. There is always some way to proceed to goals. Having documented the catalogs of losses and suffering, it is necessary to act now. The first step is to strengthen the identities and boundaries of nations. We can start by offering the status of nation, with a voice and one vote, to any culture willing to participate in a new international government.

Recognizing Catastrophes

People have been predicting catastrophes and global shifts for the past 50 years. They have identified discontinuities ranging from weather patterns to disease patterns to political upheavals and collapses of alliances. The trends they identify are unsustainable. The responses are unknown or uncertain. The predictions are uncertain. So, we need some kind of direction, as well as an understanding of catastrophes.

Many kinds of catastrophes are possible and that raises questions. What are the contributing causes? How can they be changed? Other questions include how can catastrophes be diverted or lessened or stopped? Disasters regularly arise from natural and economic systems. They are part of the process of development. Recently, the news media has identified a new catastrophe, called the 'U.S. problem' or 'the U.S. crisis,' even though it is a human crisis or possibly a global crisis. Previous catastrophes that befell other cultures, from the Mesopotamians to the Greeks and Rapa Nuins, were generally local and did not affect the remainder of humanity. Of course, the Mediterranean was changed, as was the fertile crescent and many islands. Now, we are more connected and dependent. Regional and local problems can rapidly become global.

Avoiding Catastrophes

The word catastrophe means 'down turning,' from the Greek. A catastrophe is a down-turning, literally. Most catastrophes are assumed to be fast, sudden, and manageable. But, catastrophes come in all ranges of speed, size, temporality, visibility, and combinations. Our language is poor in its terms for catastrophe. A slow catastrophe could be called a bradycatastrophe. A long-term catastrophe could be called a chronocatastrophe. A large catastrophe could be called a megacatastrophe. An unseen catastrophe could be called a cryptocatastrophe. A multi-pronged catastrophe could be called a polycatastrophe. Unfortunately, catastrophes can and do occur in many combinations. Thus the loss of species by the planet might be called a polycryptobradochronomegacatastrophe.

We know of many catastrophes that have happened in the past or distance. Such catastrophes may have included an asteroid strike that ruined the dinosaur dominance or the ecological collapse of freshwater systems, but until they happen to us, knowledge of them will not be adequate to inspire change and preparation.

Widespread poverty from gross inequities may eventually cause catastrophes, such as nuclear war or biological warfare. Richer countries will need to recognize that the poverty of others is not in their interest, especially as potential markets. Inequity may never be erased. Perhaps some inequity is good and stimulating, but gross inequity needs to be limited. Famine may come home with a vengeance, with a crisis of industrialized agriculture. But famine is not the whole problem. A related problem is the inability of people to perform on inadequate diets — to behave in a human way. This covers mental retardation, laziness and other incapacities.

Bad Circumstances
Humankind possesses incredible scientific evidence of environmental wobble, biological imbalances, and the unfitness of entire domestic species, but knowledge moves few to action. Probably nothing will be done until catastrophes become common experiences.

Perhaps good can come from catastrophes. Eric Jantsch wonders if major catastrophes on earth mean only a weeding of a garden by evolution; the loss of many lives that may permit more beautiful flowers. On the other hand, humans made the garden they way they like it. Many catastrophes would not improve our situation through chance.

Catastrophes concentrate attention on a landscape and its people and that is their benefit on human affairs. Ideally, it should not take catastrophes to precipitate corrective measures. Instead, we might resent the necessity to change. William Catton recommends that we do not indulge in resentment. Our bad present circumstances result from the innocence and hope of our ancestors; they were also the result of decisions to have babies, fires, televisions, tractors, and status. The understanding of catastrophe may let us avoid it or at least ameliorate it. Catton makes the biological analogy that die-off is a signal to overshoot, and overshoot leads to habitat damage. The agent of a post-irruption crash, for animals or humans, may be starvation, war or just behavioral stress.

If civilization collapses, the struggle back to a technological society will have greater limitations. Accessible minerals have been scattered; the gene pool has been greatly reduced. Then it may be too late. Our species may die. It is hard to image all life on earth dying. Even the worst of catastrophes would leave a simplified ecology of mosses and slugs. Weeds and invader species would prosper. Habitats would be ruled by the natural laws of ecology again. Perhaps herbivorous animals would build up large populations again. Fortunately, catastrophes do not occur in completely destructive patterns. Limited starvation occurs before total starvation. There will be uncomfortable smog before acid rains destroy crops. Some of these things could signal the necessity of immediate corrective actions.

Eric Eckholm described how economic and political pressures, which are derived ultimately from population pressures, forced farmers to intensify their efforts to increase crop production. This seems to instigate an utterly dismal cycle of population expansion, environmental deterioration and poverty: As the population expands, arable lands are used to capacity, and sometimes beyond; as the soil deteriorates, it requires more fertilizers that cause more hazardous conditions that decrease agricultural capacity; people go hungry, but the population increases, and marginal lands are used to meet increased demands, or food is imported from other lands.

Eckholm describes how the usage of marginal lands can result in a dust bowl phenomenon, when climatic conditions revert. Clouds of topsoil rolled east into cities from the dust bowl in the central United States in the 1930s. Afterwards, national conservation programs were able to restore some of the mythical fecundity, through pasteurizing, strip cropping, terracing, and contour plowing. However, current production efforts are causing greater losses of topsoil, and farmers are abandoning some of the conservation methods for economic reasons. Eckholm concludes that free market conditions encourage dangerous trends. The lesson of the last dust bowl may be forgotten until the next one.

The world food supply is in danger. Almost all the continents, except North America, now have grain deficits. And as energy and capital become dearer, the likelihood of adequately feeding the world grows remote. As topsoil loss contributes to a decline of world productivity, the higher expectations on the remaining acres may impair their productivity. This deterioration is most severe in poor countries, but its effects should concern everyone. Eckholm concludes that the United Nations must identify, analyze and marshal world resources against these trends. A scientific method will take a long time, however, and poor countries cannot wait. They must attempt a rural regeneration to stop urban drift. He interprets these trends as indicating the sinking of marginal peoples on marginal lands into a quiet helpless poverty, leading to later urban deterioration—perhaps less quiet.

The serious firewood shortage over most of the earth is also charted by Eckholm; it is considered due to population pressures on the remaining woodlands. Pressures on British woodlands in the 1300s forced people to turn to coal as a fuel source, a source then regarded as inferior. The timber famine reached Europe in the 1700s. It had existed in China and India over a thousand years before. There have been crises in clothing also, as animal hides became scarce, and then wool came into short supply as farming was expanded into grazing lands; cotton was imported from colonies, where it ruined fertile lands; now, synthetic fibers are made of petroleum products and chemicals. Each substitute required more energy to produce. Humanity has provoked a global crisis from local crises.

Fast Change
Most scientific studies have stated or implied that change cannot be fast, that people could not adjust, that social disruption would result, and that chaos would finish what ignorance and technology could not. The most serious drawback is the time of implementation. The Club of Rome claimed

a 20-year feedback lag (and it has been over forty years with mild if any implementation). *The Ecologist* cited a social inability to adapt to rapid change. Everyone assumes the time scale remaining before a collapse will be long enough for their plans to be implemented. But, these studies also propose slow, long-range plans, while warning at the same time that the earth is facing imminent, drastic change. Certainly, if their plans are implemented too slowly, and if the population or pollution doubles again, surpassing some unrecognized critical level, then there will be worse disruption. Their beliefs depend on limits to the rate of social change. As Plato recognized, it is never too late to reverse the fatal tendency towards decay, however late the hour.

It is questionable whether these time limits are absolute. For instance, war produces relatively fast, far-sighted — if wrong-headed — policies. War is actually stimulating to many individuals and often produces a national determination and purpose that peacetime does not. War unites whole peoples in a common cause. But, war is a catastrophe for most living beings and for the distribution of wealth.

There are other times when human beings face rapid and catastrophic change without chaos. Social systems can adjust if there are popular reasons. When a dam breaks, there is an emergency, and millions of people can mobilize to meet it. When the earth quakes or volcanoes erupt, when people flee from rival nations, these are emergencies, and they are met quickly.

There are millions of people starving every month in India and Africa, and others in Europe, the Pacific and the Americas. This is a catastrophe also, though it is slow, constant and quiet — perhaps because of that, it is neglected. Perhaps distance limits the ability of people to react to problems. These are emergencies and should be treated as such. Furthermore, whole species are disappearing, and whole ecosystems are wobbling, from exploitation, desertification and pollution. These are great catastrophes and much more final in character than any local one. In fact, many of the previously mentioned problems are only symptoms of these. Especially since we are part of the web of life and may be working toward our own extinction. Miners used to take canaries into the mines with them, because they are sensitive to carbon monoxide poisoning. The death of a canary was a warning. The extinction of so many species now may be such a warning to us.

The trouble with complex, self-regulating systems is that very small changes have large consequences — Reid Bryson points out that shifted rainfall patterns caused whole cultures to disappear. In some cases, where conditions, like drought, are cyclic, in the Sahel region of Africa, humans expand during the good times, only to perish when the drought returns. In other cases, human activities, such as deforestation or overgrazing of herds, can cause weather changes. We tend to approach the limits of use of some environments. We also tend to overdependence on modern high-energy methods of agriculture and on some key resources, like water.

We need to adapt consciously to slow catastrophes. The environment is changing too fast for genetic adaptation, so our change will have to be psychological and social. Social changes can occur very rapidly when the time is right for them. Oil-producing nations, for instance, became financial equals of industrial countries within months.

Pressures are building for radical change. Change can intensify and accelerate. Small changes in different weather patterns can lead to dramatic changes in climate. Changes in climate can lead to dramatic changes in the species composition of ecosystems, as well as to human food production and housing costs.

Applying a Disaster Psychology
Catastrophe has its own psychology. Humankind possesses incredible scientific evidence of environmental wobble, biological imbalances, and the unfitness of many domestic species, but knowledge moves few to action. Immediate foreknowledge provokes a greater response than indefinite expectations: hurricanes occur periodically, but not always in the next three hours. This concentrates our attention forcefully. The psychological effects of a hurricane—fear of suffering and dread of loss, accompanied by exhilaration—have admirable 'side-effects;' the definiteness of danger and the immediacy arouse people to great heights of cooperation. People react admirably to catastrophe. They choose sensible directions and agree to practical expedients. There are times when human beings face rapid and catastrophic change without chaos. They can adjust if there are popular reasons.

Environmental deterioration proceeds so slowly that the change is invisible, that is, until a catastrophic threshold is crossed. Those catastrophes, in places from the Tigris and Euphrates to northern Africa, seem to be long-term. . In other cases, human activities, such as deforestation or overgrazing of herds, can cause climatic changes. The scale and rate makes our situation seem natural, but that is because it has been a slow catastrophe, just now approaching the threshold.

Probably nothing will be done until catastrophes become common experiences. As choices become more important, more urgent, errors are more disastrous. It should not take catastrophes to precipitate corrective measures.

But, humans should not adapt to these catastrophes. As Rene Dubos has pointed out, humanity is enormously adaptable and resilient. We could probably survive almost any physical or social conditions by adjusting to them; for example, overcrowding and smog are the norm in some areas. Hans Selye suggested that an organism is in more danger from its adaptive reactions than from external agencies. Adaptation has its own dangers. We might become less humane, less creative, and less concerned with starvation, suffering, crowding, or destruction. Our goal should not be to survive under any conditions, however difficult and unpleasant. Our goal should be to create an optimum life in an optimum environment. Perhaps this goal needs to be framed first in a humanistic or religious framework.

We do not know where to look for meaning in a cultural world, plagued with dispossessed people. Nothing is meaningful if civilization

goes mad changing whole ecosystems, as Paul Shepard suggests it is.
The adaptational environment is too fast for genetic adaptation. We have adapted, with our culture, from nomadic hunter to agriculturalist to urban chemist in a short time. Now, the speed of technology seems to be too fast for cultural adaptation. We need to understand how to react, to control the tempo or slow down.

If we could precipitate a disaster psychology for slow environmental or cultural catastrophes, then the priorities and motives of people might be changed. But how should everyone be convinced that there is a crisis? The change is so slow: Fewer eagles, fewer salmon, more people, more beverage cans. The causes are so complex, and responsibility is so difficult to assign.

Unfortunately, a history or theory of catastrophes does not engage us. People need to feel situations before they act, and people will need to feel themselves as part of a delicate web of relationships before they act with ecological wisdom, as once they had to feel that the earth was round by going around it, as lately they had to see that the earth was an oasis in space by leaving it briefly.

Taking Immediate Action

A transformation of world order is necessary before the next large catastrophic event. But perhaps, as many have said, the dreadful has already happened—in Hiroshima, Vietnam, Iraq, Africa, and lesser known places—and the cultural transformation is already awakening or has already finished. Most cultural transformations are invisible.

Maybe the gloomy forecasts are right. But, it is the function of a Cassandra to be always wrong. To be successful, Cassandra always has to be willing to be proved wrong. If she is disbelieved and her warnings go unheeded, then her prognostications may come true. But, if she is given credence, steps are taken and policies are changed to falsify her warnings.

Political response is not enough to change the system. Many problems are social or cultural; many are personal. As people change their attitudes, social and political changes can be made. This could take generations. A violent revolution could occur within weeks.

The future is said to be in our hands, but our hands are in the cookie jar or on our genitals, and we are afraid to let go to touch something different. Perhaps we are just insecure. Disaster is not inevitable, unless we refuse to change our ways, in which case disaster is inevitable. So, we need to change. The change is going to discomfort many, but not as many who are being discomforted by not changing directions.

How long do we have before we have to change? A decade? A hundred years? When should we start? Last minute? Last week? Next month? How should we acquire urgency? Or the political will to change or the wisdom to change in better ways? What are our long-term interests? Is it necessary to declare a state of emergency? Another alternative? Not a war against nature or injustice, which is a tired metaphor, but just a rational response to a long, slow catastrophe. The response has to be immediate action.

Taking Coordinated Action on Three Levels

After adopting an attitude towards catastrophe and starting to take action immediately, people, nations and a global framework will continue to act on those three levels. All three are necessary, but perhaps the individual is the most critical and important, if not to make changes, then to stimulate change for others and at higher levels.

This action has to include a survey of the state of the world, of all ecological and social systems. We have to make radical changes in the economic and political systems that arrange, produce and distribute wealth. We have to create a framework for world order without making the mistake of a gigantic global dictatorship, even if it is for the good of equality and environmental health. An order has to be based on bioregional models where nations are based on ethic populations in specific regions, where the operation is of interlocking hierarchical systems.

Acting as an Individual/Community

The redesignation of cultural nations does not involve a major revolution for most people. Most people prefer to remain in their native culture, with a geography they fit into, an economic system that reflects their values, and a religion that explains their image of place. Revolution is a false dilemma, it does not reflect possibility of thousands of individual actions on farms, factories, and families, all at local levels. Individuals can control their lives. They can make choices to be self-reliant or to limit their impact on their supporting environment. Margaret Mead noted that this is how anything has ever been done, by the actions of committed individuals.

Of course, small steps can work. Little things matter, if the scale of the repetition is large enough, that is, if others do it also. Individuals can try to question things that seem to be automatic behaviors, such as wanting children or requiring personal transportation.

Live Frugally & Eat Well. A first step for an individual can be to live frugally, as suggested by Henryk Skolimowski. You can reduce your role as a consumer by not seeking meaning in the acquisition of things. That would result in fewer clothes, fewer tools, fewer toys, and fewer luxuries, which would force the economy to change from obsolescence and growth to development and permanence. That does not imply being impoverished, just fewer luxuries.

Simplicity can continue with the diet, by eating lower on the food chain, with more vegetables and fruits, and by getting local food, which keeps the money circulating in your neighborhood. Reducing the consumption of meat will reduce the demand for meat and cattle, which will reduce the use of rangelands and resources for these animals.

An individual can live without many personally-owned things, such as a computer, table saw, car, or lawn mower. Many of these things can be rented for short periods of time, if they cannot be shared among the neighborhood. If you live in a city, use public transportation or cars for hire.

Make Homes, Personal Possessions & Things. In many cultures, individuals assembled their own houses. That may not be possible if the house is complex, but any individual can participate in finishing and personalizing their house. A house is often the largest item people will build or buy. There are many ways to reduce the impacts of homes, for instance, build in a place where it will not cover fertile soil or interfere with water cycles. New houses are a major source of habitat loss.

Reduce the immediate impact on its setting. A big, new tract house with a lawn that demands water, pesticides and a fuel-consuming power-mower, can be avoided. Individuals can make their houses more efficient and naturalize their settings, with native plants in a sensible arrangement. If you have a yard, use natural plantings that require only natural water from rainfall. Arrange them sensibly around your house to cool or protect the house. Make things that you need to keep your shelter useful. Compost organic waste to renew the soil.

Reduce the size of the house. Houses have been ballooning in size, so that recently in the US, the average house is over 3000 square feet. All that extra space needs to be heated and illuminated, mostly by burning fossil fuels, and all that extra material needs to be extracted from the earth, then manufactured and transported. For the US, 500-600 square feet per person might be a good average.

Reduce Consumption of Materials & Space. Avoid buying things that cannot be recycled or reused easily. Avoid buying cheap things that have a useful life of less than a year. Avoid buying things that you do not need. Avoid paying for them with credit. Use less paper or energy at work. Reuse paper and envelopes.

Other needs, such as pets, can be shared, also. Michael Fox has suggested that shared pets would benefit many pets and reduce the number of unwanted animals — the millions of unwanted animals that are abandoned and then often disrupt wild or conservation areas. It may be necessary to destroy large numbers of pets that have gone feral. It may be necessary to reduce the number of pets, through licensing, especially emphasizing working animals such as guide dogs.

Reduce Consumption of Fossil Fuels & Reduce Emissions. Walk or bicycle if you can. Take a bus or train if you can. Get a hybrid or hydrogen-powered car if you can; otherwise try to buy a more efficient combustion engine, some of which can get 50-70 miles per gallon. Share your car with others in a carpool or pooled shopping trips, or share their cars. Take care of the car by getting it tuned and keeping tires inflated. Drive it more slowly.

Individuals can work to make changes, such as reducing the number of cars and roads, especially in roadless areas. Perhaps the number of cars could be restricted by a modified lottery system that allowed emergency vehicles and special needs vehicles and then randomized the remainder below an absolute number, which would be related to the number of roads and their condition, including crowding. Individuals could encourage a road-building moratorium; fix up the old infrastructure or replan what roads to be discontinued or changed into areas for housing or commerce.

Reduce Material Waste. Reduce unnecessary waste in your home. Do not heat it or cool it beyond a reasonable amount; if you want to be really cool, for instance, consider moving closer to a planetary pole. Use natural kinds of change such as open windows or evaporative coolers. Use natural light or mechanical nonelectric devices to open cans.

Reduce toxic substances inside the house. There are many potentially hazardous chemicals and chemical combinations, even with substances that are benign on their own but toxic in combination. Recycle things like aluminum, if you cannot avoid buying aluminum containers; recycle glass, batteries, and newspapers, if you cannot avoid them by reading on a computer or discussing things with neighbors.

Be Healthy. An individual can eat well and exercise regularly. Exercise can be a part of everyday living if you can walk to work or bicycle to visit friends. Reduce the amount of time sitting, especially in front of computers or televisions. Televisions and computers can be important parts of our education and communication, but they should not be the only part or the most important.

Develop Your Self. As an individual, you can learn something new about your location, for instance about many of the invisible parts of your environment, from insects to elemental cycles. Develop yourself, before having children or caring for children.

Insist in pursuing a form of livelihood that is interesting and rewarding. The most important loss, according to Robert Reich, is the loss of the individual's power over their livelihood. Fewer people have the right or opportunity to choose their work, to make a living, to participate materially and meaningfully in society. They are constrained to taking opportunities present in the system, which is indifferent to people, other than as disposable, replaceable or surplus cogs.

Play. Individuals can engage in meaningless, unproductive activity. Do something for fun, without regard to its usefulness or rewards. Make something and take it apart. Play. Play is imaginative experience. Play is not limited by arbitrary rules and economic goals.

Play is the method of learning for most juvenile animals and a means of enjoyment for many adult animals. For humans, play is imaginative experience, entered into freely. Much human activity is play, in place in a community. Even science and philosophy are forms of play, as attempts to solve the puzzles of existence. The state of play is whole and simultaneous. The rich flowering of human nature is possible only when the constraints of need are replaced by leisure and abundance.

Participate in Community & Culture. Individuals need to participate in their communities. They can join conservation groups and volunteer to help with specific projects, such as saving bird or bat habitat. Work with the county or state on long-term plans for beautification or change. Planting trees works well for carbon sequestration, as well as beautifying an area and filtering pollution.

Work to Guide Nation & Corporations. Try to limit the power of corporations; work to revoke or amend their charters. Redefining corporations from fictional individuals to real, responsible collectives would make them more responsive to their effects and damages. Work to equalize things. For instance, try to tie politicians salaries to the average working wage, which would certainly be an incentive for them to increase the average; and, try to tie their generous retirements to standard programs, such as Social Security in the U.S.

Make sure that political representation is more equal, with less gerrymandering by politicians for their own benefit. The Electoral College, for instance, was a good idea for a time when news was slow and people were less informed. There is no excuse for ignorance of the voters, now, so direct elections should be more effective. Try to get the constitution amended to reflect the value of wild nature and its automatic services that are incredibly expensive for us to duplicate.

Seek to Limit the Domination of Government & Corporations. Individuals can try to make sure that the size of corporations and governments fits the size of your community. Get commitments to invest locally to keep symbolic wealth cycling locally. Take community action to describe limits and restraints on corporations. Take legal action to correct corporate irresponsibility. Take personal action to educate others about elegant simplicity. Take political action to get representatives to address these issues, from local safety to international coordination.

Protest unhealthy actions. Write letters or make conversations about concerns. Start a dialogue with your neighbors. Take direct actions against harmful behaviors, which may be personal, community, or corporate. Get your community to be weapon free, not only from large weapons of national defense but also from personal weapons larger than knives. Get the community to observe itself.

Reduce Conflicts with Consensus Conversation & Compromise. Individual steps are incredibly simple. Start by listening. Accept what other people say and try to understand a different perspective. Many conflicts result from misunderstanding, and many misunderstandings can be resolved through communication.

Dualisms Conversations & Consensus. Many archaic cultures had formal paths to resolve conflicts, through conversations and consensus mediation. The principles of small cultures are similar: All decisions were made by consensus in which everyone participated; chiefs were not coercive rulers, but teachers and leaders with specific duties limited to their realm — medicine, war, or ceremonies for example. Cooperation and consensus, as opposed to competition and individual exaltation, permitted planning to remain informal.

Government by local meeting assumes the common sense and wisdom of the common person in an open exchange of belief and need. It requires trust and esteem. Often this kind of involvement takes more time than just voting annually or having one person make most decisions. The effect of presenting a problem before an American Indian council was to slow down response by passing it to the entire constituency and getting a consensus. This ensured due consideration. Living generations are responsible for limiting their actions within a reasonable framework of cost and irreversible change. The standard requires conversation and some consensus about the limits, which are never exact. Ecological value has to be balanced with socially optimal resource allocations that consider past and future generations as well.

Communicate Common Sense. Individuals can try to get other people to agree to be use more common sense. Try to get legislation changed to favor conservative measures. For instance, reducing the number of cars may be difficult. After all, the entire might and weight of advertising dollars is given to showing large vehicles ravaging the natural environment for fun — and it is fun to drive large bush-ripping, flower-crushing vehicles across the streams and the entire landscape, make no mistake — it just results in the suffering of other people or living beings.

Seek to have federal laws requiring recycling, as part of a larger use program. Saving humanity, or the planet, is ultimately a political task, and people do have power. Citizens who were concerned about the removal of ozone-depleting chemicals from cosmetic products, or the efforts to ban unnecessary herbicides and pesticides, convinced others that these were important topics, requiring legislation and marketing awareness. As the marketplace has taken over a larger percentage of the public arena, people have more clout than ever, because the marketplace is accessible to everyone, and it is vulnerable to fads as well as to safety issues. Businesses are vulnerable to much smaller trigger effects. To change business emphasis, it may take only a change in profit margin of a percent or two. Mobilization, through communities, television, or the internet, has been made easier. Everyone wants to live in a healthy environment, and to provide one for the following generations.

Volunteer for Your Community. Volunteering is just the formal recognition of the required unavoidable participation in one's native society. It is formal because many people are estranged from the requirements of a modern culture or have forgotten how it works, as a result of alienation. Archaic peoples knew when to help others on projects. Modern people often have the impulse, which is reflected in the many volunteer organizations, but the effort is often disconnected, uneven or inadequate. This service is common sense. You learn from the community that supports you with help and love. Each person should spend from two to four hours per week on community projects, from holding office to being a companion to the young or elderly.

Serve Your Nation. Everyone should agree to serve their nation for a two-year period, by working on educational, security, maintenance, or emergency projects. The infrastructure of a nation, its roads, bridges, and canals, often need more maintenance than can be hired out. Special amenities, such as parks or low-cost housing, are often not considered profitable enough to design or build. Although emergencies rarely visit a community more often than once every twenty years, across a nation, emergencies are daily events and critical enough to need extra labor to resolve.

Volunteering would also reduce the need for standing armies or drafts for military conflicts. This kind of service can be put to work on restoration projects, from houses and neighborhoods to conservation areas and wilderness areas.

Many other activities can serve a nation. You can also serve the nation by sponsoring laws that raise a minimum wage or issue coupons for minimum support for food, education, and health-care. Writing books and creating designs is also a service.

Assess Your Self-Performance. One can always improve one's behavior. Sometimes people will consume too much or eat unhealthy foods or drink too much. The act of self-assessment can help people reduce self-destructive or harmful behavior.

It is important to keep things in perspective. Whether or not we buy a second car makes a big difference; washing out plastic grocery bags makes a smaller difference.

Question Things. The Union of Concerned Scientists has a handy rule-of-thumb for consuming: Ask yourself, "How big is it?" Buying an energy-efficient refrigerator reflects more on conservation than replacing the light-bulb inside it. Buying juices in cardboard containers has less impact than recycling aluminum cans. Another good test is to ask more questions: "What is this purchase supporting? Is it efficient production? Does it work towards my personal needs or goals?" Driving a fuel-efficient car or choosing nonbleached toilet paper is a decision that will be noted by the manufacturer, the distributor, the retailer and the market.

Act on the Answers to the Questions to Explore Changes. Asking questions is a good start, but then you should take actions based on the facts that the questions have uncovered. Actions have some effects elsewhere, even if they are not immediately apparent. Explore changes in yourself as a result of your actions. Adjust your behavior based on what you see. Maybe the actions will not have exactly the effects that you want, but you will be more likely to understand how these actions mesh with others.

Community Steps

People live in communities. The community is responsible for the health and activities of individuals. Common values are reflected on the community level, since they are imbibed by individuals in the community. The community has distinct actions that can be taken for human culture to get to good places.

Defining Itself as a Community or State. Human community organization is a cultural universal. Communities are self-making. A community has a special image and set of special beliefs related to place and history, such that they are unique and cannot simply accept many of the beliefs of other communities.

A community implies that the experiencers share ways of experiencing or the same experiences. This sharing enables an individual to go beyond a finite view, to see the embedded community as one of many ways of relating the self to the universe.

Setting the Goals of a Community. Unconsciously or consciously, a community sets goals related to its style and size. The science of ecology to identify those goals, especially regarding impacts on wild ecosystems. Ecological knowledge forms the basis of the normative aspects of living together, that is, ethics, and the maintenance of the affairs of a community, that is, economics and politics. There can be no separation of politics and ecology. Every economic or political act has ecological consequences and every ecological decision is an economic or political demand for control over use of the environment.

Implementing Ecological & Cultural Goals. Local goals are appropriate for watersheds and habitats. Educational and social goals also originate at the local level. These goals are not exhaustive:
- Zone wild ecosystems first
- Measure the productivity of all ecosystems to contribute to a regional or global inventory, from which to make intelligent decisions
- Preserve all special areas, such as heritage land or old-growth areas
- Protect the long-term health, integrity, and ecological balance of ecosystems, that is, the ecological and evolutionary processes that make ecosystems
- Work with indigenous peoples to get control for them of their lands, which are often an important part of their culture
- Protect local water and air sheds
- Provide habitat for all species, since all species contribute to the functioning system, even agents of disease
- Manage commercial land for permanence and sustainability
- Provide exclusive areas for recreation and aesthetic appreciation
- Achieve true multiple use of land by strict regulation and rationing
- Set up land trusts for intergenerational protection
- Protect the health of human communities

- Open communications with all groups working in the region
- Develop paths for public participation in land use decisions
- Broaden local economies from resource extraction to invention and specialization
- Set up cooperatives to refer and share work
- Combine forest and agricultural crops where appropriate (through tree crops or permaculture)
- Diversify local institutions that deal with ecosystems
- Encourage survival of small, ecosystem-based communities
- Increase manufacturing efficiencies above 75 percent (compared to a 50% world average, 67% in Japan, and 40% in Thailand)
- Promote the full use of ecosystem products; support small-scale businesses that produce new, high-quality products
- Educate all people to feel their connections to their place, because, until they feel them, they will not act ethically or ecologically
- Educate people to realize that long-term sustainability requires healthy places, and that protecting places protects jobs
- Educate people in the science, art, and techniques of place (literacy, numeracy, ecolacy, and imagacy)
- Develop criteria for the ecologically responsible use of places, as well as standards for certifying practices and products
- Work to incorporate findings from practical and theoretical sciences into a unified knowledge of places
- Resacralize places — desacralization is a defense mechanism against loss of meaning — by emphasizing the meaning of wild places
- Base the size and kind of a culture on the limits of the ecological place.

These goals, and others, need to be discussed and modified in a community context, through discussions and meetings.

Convening a Constitutional Meeting. Starting is simple; a community or group of communities in an area, for example in Eastern Washington and Northern Idaho in the U.S., could convene a constitutional meeting. The constitutional convention would work out a new government, possibly more radical than the 1972 Montana constitution, which required a local government review process, which mandated real change in the form of government, although few changes took place. The convention may suggest new boundaries of areas, according to watersheds or other boundaries. An example is in the Washington State statute of 1967.

Because of the differences in size, and the need for even the smallest community to be enfranchised, it may be useful to adopt a floterial district system as in Idaho. A representative could represent other communities.

The convention could draw on the cultural depth of the region. Goals that have been only thoughts or hopes could become explicit expressions. This kind of change would depend on the flexibility of the federal government to accept the reformation of some of its states or regions. If the communities decided, they could create a new nation.

The new nation could join the Unrepresented Nations and Peoples

Organization (UNPO), which promotes the respect of human rights of all peoples through nonviolent tolerance and self-determination. UNPO offers an international forum for nations and peoples whose causes are not addressed by the existing UN or by existing bodies. UNPO provides assistance to 26 groups, representing 50 million people, including the peoples of Mari and Tibet. The new nation could give recognition by other unrepresented nations, such as Scotland or the Karen. The nation could apply for recognition as an independent nation through the United Nations or other new international organization — perhaps a Global Union.

Balancing Budgets for Community or State. A community needs a budget to limit its wealth to actual quantities. A budget generally refers to a list of all planned expenses and revenues. A budget is an important concept in microeconomics, which uses a budget line to illustrate the trade-offs between two or more directions of development.

The budget of a government is a summary or plan of the intended revenues and expenditures of that government. A budget is usually prepared by a separate department and then submitted to the legislature for consideration, changes and approval. Any budget should be required to be balanced in the long run.

As the technology of money and banking continues to make spending easier, individuals, families, companies, and governments need to adopt more effective budgeting methods, strategies, and tools to balance their outflows to their inflows and to avoid deficit, or debt-based, spending.

Annual or multi-year budgets could be replaced or complemented by monthly forecasts or rolling forecasts. Monthly forecasts provide more up-to-date financial plans. Many large businesses reforecast their original budgets on a quarterly basis. As months pass, the actual income achieved and expenses incurred can be compared to the budget and forecasts. Variances between these financial plans and actual delivery can then be analyzed to provide information that can improve performance.

Income for Community or State

Income is the form of usable or tradable wealth that is used by the community to pay for public expenses and amenities. As Herman Daly and John Cobb point out, the means of raising community funds would also be the means of attaining personal and community goals. Taxes could be ways of internalizing the costs of the community and some artificial constructs, such as corporations.

Taxes for Community or State. This term 'tax' in its most extended sense includes all contributions imposed by the government upon individuals for the services of the state, by whatever name taxes are called, whether it is tribute, tithe, talliage, impost, duty, gabel, custom, subsidy, aid, supply, excise, or other name. Taxes are any charge of money or property imposed by a government upon individuals or entities that are within the government's authority to assess such charges. This term, however, generally does not include charges imposed in exchange for the provision of specific goods or services, such as bridge tolls or sanitation fees. Although most modern taxes are levied on the basis of economic measurements such as income, consumption, property, and wealth, some governments also impose excise taxes or other taxes.

Traditional Kinds of Tax. Many traditional taxes require complex schemes to avoid cascading taxes or the regressive unfairness of some taxes. Many taxes were *ad hoc* additions to a tax code to add income from new or overlooked profitable activities.

State of Being Taxes, or Title Taxes. An example of a 'state of being' tax is an *ad valorem* property tax, which is not an excise tax, but which may be imposed on the property or the person who owns that property at a certain moment of time, for example, on July 1st of each year based on the state of title at that given moment. The 'state of title,' state of ownership, or property by reason of its ownership, is what is being taxed. The next year, on July 1st, another such tax is imposed again in the same way on the same property and person, even though there has been no 'change' or intervening 'event.' The amount of the tax may change from year to year, based on the change in the value of the property or a change in the tax rate, or both, but those are separate issues governing how the tax is computed. What is being taxed, fundamentally, is the state of title, which is not an 'event' but instead a 'state of being.'

Event Taxes. For purposes of the U.S. constitution, for example, an excise is essentially any indirect tax, or 'event' tax. In the constitutional sense, an excise includes gift taxes, estate taxes, payroll taxes, sales taxes, miscellaneous excise taxes, and income taxes on any income other than income from property — in short, any tax that is not a direct tax. Excise taxes are taxes paid when purchases are made on a specific good, such as gasoline, which is often one of the major components of an excise program. Excise taxes are usually included in the price of the product. There are also excise taxes on activities, such as on wagering or on highway usage by trucks. An excise tax can have several general excise tax programs.

Excise duties usually have one of two purposes: To raise revenue or to discourage a specific behavior. Taxes such as those on sales of fuel, alcohol and tobacco are often justified on both grounds. Some economists suggest that the optimal revenue-raising taxes should be levied on the sales of items having an inelastic demand, while behavior-altering taxes should be levied where the demand is elastic.

Capitation Tax. The capitation tax or "head" tax is an assessment levied by the government upon a person at a fixed rate regardless of income or worth. A poll tax is an example of a capitation tax. Originally used as a duty on the importation of convicts and slaves, it was later used to discourage the poor from voting.

Severance Taxes. A severance tax, on items that are 'severed' from their context, would have the effect of limiting the use of nonrenewable resources, such as coal or oil, as well as slowly renewable resources, such as forests.

Income Taxes. In 1913, the Sixteenth Amendment to the U.S. Constitution was ratified. It empowered Congress to tax "incomes, from whatever source derived, without apportionment among the several States, and without regard to any census or enumeration." The Internal Revenue Code is today embodied as Title 26 of the United States Code and is a direct descendant of the income tax act passed in 1913. While some U.S. states, such as Nevada or Florida, do not have an income tax, all residents and citizens of the United States are subject to a federal income tax, which requires a minimum level of income.

Sales Taxes. A sales tax is a tax on the consumption of goods or services. Normally, it is a certain percentage that is added onto the price of a good or service when it is purchased. To avoid double taxation, or 'cascading' taxes, where an item is taxed from design and production to retail sale and resale, a sales tax should charged ideally only once on any one item. A retail sales tax is charged only on retail transactions, not on businesses buying raw materials for production or on businesses reselling finished goods.

A related type of sales tax is the *value-added tax* (VAT). All sales, retail and wholesale, are taxed in this scheme. To avoid multiple or cascading tax, every business is refunded the amount of VAT remitted by their suppliers. Most nations, especially European nations such as Norway or Ireland, have sales taxes or VATs at the national or local level.

Effects of Traditional Taxes
The combination of these taxes often makes them regressive. In terms of fairness, sales taxes are generally regressive, that is, poorer people tend to pay a greater percentage of their income on sales taxes than richer people, because they generally tend to spend a far higher percentage of their income for food, clothing, shelter, and medical care. In some locations, items such as food, clothing, or prescription drugs are exempt from sales taxes, ostensibly to alleviate the burden on the poor. Some of these exemptions, such as exemptions for clothing or prescription drugs, may actually have the opposite effect. Some of these taxes discourage the efforts that they tax. For instance, a tax on value added by labor to commodities could discourage labor itself.

New Kinds of Taxes for Community or State
When Arthur C. Pigou introduced the concept of 'taxing bads' in the early 1900s, economists and politicians discussed the idea, but its use was limited. Environmental taxes are functionally nonexistent in the current tax code of the United States. A few such taxes have been proposed, most recently a BTU tax during the Clinton presidency, but the defeat seems to have discouraged any kind of environmental tax since then.

Taxing 'bads' is essentially a 'tax shift.' Taxes are reduced on things that should be to be encouraged, such as work or savings. The loss of revenue is compensated by taxes on 'bads,' that is, on things to be discouraged, such as pollution and waste. A tax shift could help mitigate the impact of pollution charges on some businesses; it would remove taxes that discriminate against low-income families. A tax shift might also improve general fiscal health, as well as over-all environmental health.

Rather than use old labels, which were the result of *ad hoc* additions to many tax codes, it might be simpler to present them as what they are taxing, that is what kinds of income we want and what kinds of behavior we want to encourage. These kinds of new taxes can be put under four categories: Use, Loss, Adjustment—or the misplacement of resources, such as pollution—and Distribution.

Use Taxes for Community or State. A use tax is similar to many consumption taxes or a few severance taxes. These taxes take a percentage for a service for any resource. A use tax would have the effect of limiting the use of nonrenewable resources, such as coal or oil, as well as the use of slowly renewable resources, such as forests. The rate of the tax would be related to the scale of the economy, as well as to the carrying capacity of the ecological support system.

Tax collectors would monitor points of entry, from wellheads to forests, to ensure that the tax would be fair, and that it would be paid. The tax would be easier to collect and harder to avoid than income taxes. It could be, and probably would be, included in the cost of any commodity that used the resource. Few commodities do not require some resource.

Air Use Tax for Community or State. Taxes on output or on polluting inputs are called presumptive because their target is the pollution presumed to be associated with the activity. A combination of the two types, those that reduce output and those that reduce emissions per unit of output, could replace emission fees and reduce monitoring requirements and their prohibitive costs. The taxation of fuel use can be a powerful, indirect instrument for controlling air pollution because of the direct connection between fuel use and emissions. This is not quite the same as taxing the pollution; in this case the goods are taxed to address the pollution.

Indirect instruments, such as a tax on air, are designed to reduce the scale of output, as important complementary measures in a program of cost-effective pollution control. Air is used in almost every operation of an industry or bureaucracy. Clean air is used for cooling and to provide oxygen for the occupants of buildings.

Water Use Tax for Community. Water is such a good solvent, it can be used to enhance many kinds of processes. Water is used in almost every operation of an industry or bureaucracy. It also is used for cooling and for drinking.

Bottled water consumption, which has more than doubled globally in the last six years since 1999, is a natural resource that is greatly affecting the world's ecosystem, according to a new U.S. study. "Even in areas where tap water is safe to drink, demand for bottled water is increasing, producing unnecessary garbage and consuming vast quantities of energy," according to Emily Arnold, author of the study published by the Earth Policy Institute, a

Washington-based environmental group. A water use tax would encourage efficient use of water by making waste too expensive.

Land Use Tax For Community. Land use tax is a base tax on land, which would encourage efficient use of the land, as opposed to, for instance, building taxes, which if higher than tax on land allows land to lie idle or buildings to deteriorate. It would reduce runaway and harmful speculation on land. Users fees would have to be high enough to cover the costs of the services provided, according to Daly and Cobb. Cities that increased land taxes and decreased building taxes have encouraged better building programs.

This tax would tax the use of soils by agriculture or forestry, especially rapid-use forestry with short rotations (or perhaps on less than 500-year cycles). Agricultural land should also be taxed on its unimproved value, according to Daly and Cobb. The taxes would be local rather than national. Improved agriculture, such as regenerative, would prove to be more efficient as a result of the higher initial cost and the lower taxes on improvements or profits.

Element Use Tax for Community. Under the Clean Air Act, the U.S. Environmental Protection Agency (EPA) sets national standards for ambient air quality designed to protect public health and welfare. The EPA defines acceptable levels for six 'criteria' elements that are air pollutants: Sulfur dioxide (SO_2), nitrogen oxides (NO_x), ozone, particulate matter, carbon monoxide (CO), and lead. Along with emissions from natural sources, emissions of air pollutants from stationary sources, such as industrial facilities and commercial operations, and mobile sources, such as automobiles, trains, and airplanes, contribute to the ambient levels of those pollutants.

Some U.S. states levy a 'severance tax' on mineral production because these assets are nonrenewable. Once the minerals are produced, or 'severed,' from the ground, they are gone forever. Since many of these assets, such as coal and oil, come from federal lands in the U.S., which are owned by all the citizens of the states and the nation, citizens are entitled to a fair rate of return on the assets. An element tax could capture the value of these assets for current and future generations of citizens.

Carbon Tax for Community. Carbon appears in every living cell; life on earth is carbon based. However, out of place, carbon can poison life or cause changes to the climate of the planet. Carbon dioxide is recognized as a greenhouse gas, and increasing levels in the atmosphere are linked to global warming and climate change.

A carbon footprint is a measure of the amount of carbon dioxide emitted through the combustion of fossil fuels. In the case of an organization, business or enterprise, a carbon footprint is part of their everyday operations, as it is for an individual or household. A carbon footprint is often expressed as tons of carbon dioxide or tons of carbon emitted, usually on an annual basis.

A carbon tax would link the effects of burning carbon-containing fuels to the cost of repairing the damages from burning. A carbon tax could greatly offset other tax rates, as well as reduce energy consumption, especially fossil fuel consumption, and address climate change. Although there would probably be a small reduction in measures of wealth, such as the GNP or ISEW, there would be offsetting tax reductions elsewhere in the economy; and some of the money would go toward improving efficiency.

Displacing these old taxes with use taxes would improve the productivity of the economy. A high carbon tax, coupled with reduced tax rates on income and profits could generate a significant gain for each dollar of tax shifted. The gain would come not only from improved economic efficiency but from reduced investment in infrastructure, in reduced operating costs due to higher energy efficiency, and in reduced environmental damage.

Nitrogen Tax for Community. Nitrogen is a critical element of life; it is a major part of the atmosphere. In the form of nitrite, nitrate and ammonia, nitrogen enters watershed, oceans and streams from fertilizers, animal wastes and decomposing organic matter. If nitrogen levels increase, algae increases dramatically. Nitrogen oxides (NO_x) usually enter the air as the result of high-temperature combustion processes such as those in automobiles and power plants.

One way to help control NO_x would be to tax emissions from stationary sources. For example, firms might adopt currently available abatement techniques whose capitalized costs are lower than the tax they would otherwise pay. If a regional allowance trading program was put into place, the community could tax only the stationary sources of NO_x that do not participate in the program.

Sulfur Tax for Community. Sulfur is also an element required by living beings, and it also appears in the atmosphere. Sulfur dioxide belongs to the family of sulfur oxide gases formed during the burning of fuel containing sulfur and during the operation of metal smelting and other industrial processes. Exposure to high concentrations of SO_2 may promote respiratory illnesses or aggravate cardiovascular disease. In addition, SO_2 and NO_x emissions are considered the main contributors to acid rain, which can degrade surface waters, damage forests and crops, and accelerate the corrosion of buildings.

One option is to tax emissions of SO_2 from stationary sources not already covered under the acid rain program. Marketplace strategies can be coupled with regulatory strategies.

Other Element Taxes for Community. Phosphorus is a relatively rare element. Isaac Asimov calls it the bottleneck of life, a good example of Liebig's law of the minimum, which states essentially that something has to be in least supply. In the case of life, it is phosphorus. Phosphorus is a necessary constituent of protoplasm, but most of it is locked up in phosphates in rock, which is weathered and 'escapes' to the oceans, which act as a sink. Sea birds play a large part in returning it to land. In fact, guano deposits are mined for fertilizers. Harvesting fish for food returns some to land, but little effort is made to recycle it. A community tax on phosphorus would encourage more of it to be recycled.

Oxygen is a major component of the earth's crust. It is a significant component of the atmosphere. For living beings, it allows energy to be released as part of their respiration. In the atmosphere, oxygen can be converted by natural processes, such as sunlight or lightning, to ozone. Ozone is not emitted directly into the air but is formed by the reaction of volatile organic compounds (VOCs) and nitrogen oxides (NO_x) in the presence of heat and sunlight. Ozone occurs naturally in the upper atmosphere and provides a protective layer there. At ground level, however, ozone is a prime ingredient of smog. Short-term exposures, from one to three hours, to ambient ozone concentrations have been linked to increased hospital admissions and emergency room visits for respiratory ailments. Ground-level ozone has remained a pervasive pollution problem in many areas. Rather than directly tax oxygen use, communities might want to tax the VOCs or chemicals that interfere with ozone.

The advantage of a broad-based tax on VOCs is that it would affect large and small sources of the compounds. A disadvantage of such a broad-based tax is that it may be regressive. To the extent that the tax raises the prices of consumer goods, it may take up a larger share of household income for low-income consumers than for higher-income consumers.

Species Use Taxes for Community. Tax would be charged on the taking of members of any species; this tax is not the same as a hunting or fishing license, which is required to test knowledge and dictate that only a certain number can be taken. Many of these species are slowly renewable or functionally nonrenewable. This tax would work to encourage preservation, take-limits, and take-efficiency. This would tax the use of virgin species for commerce. Species, as well as fresh air and water, are the interest of an ecosystem; the ecosystem itself, on life-bearing land, is the capital. If the capital is allowed to be used, the taxes on the land should be very high. Using the interest is a more sustainable practice.

Loss Taxes for Community
A loss tax is a tax on losses from the capital base, that is, it is a tax on the destruction of resources, not just their use or on their negative impacts. It is similar to a dispossession tax or a capital depletion tax.

Resources can be sorted into one of three types of resources: (1) fastly accruable, that is, renewable in economic terms over a short time horizon; (2) slowly accruable, which are basically nonrenewable within a human lifetime; and (3) and slowly dispersed, which are really nonrenewable in most economic frames, but are actually renewable in geological time, which is rarely considered. Most of the wealth used by modern economies is nonrenewable. These resources are limited, interrelated and distributed unevenly. Forests are a special problem; although trees can grow to a good size in 30-40 years, forest ecosystems may take 300-600 years to develop and then last for thousands of years.

Oil, coal, peat, and some woods are functionally nonrenewable. Geological time periods are required to produce them. Mineral reserves may be understated, but, they may be located so far and so deep that it would cost more energy to extract them and move them than they are worth, unless we used a 'renewable' energy source—the sun.

Slow accrual and slow dispersal resources should be equally available to all cultures. It is impossible to sustain any quantitative arguments about resources and population pressure on them without a comprehensive overview. Demands on food, fertilizer, energy, and metals are related inseparably. Organic and inorganic assets need to be assessed together. Population carrying capacity cannot be formulated until both resources have been quantified.

This kind of loss tax is on things or processes that interfere with other things and processes, things that cause runaway feedback or the destruction of cycles, things in other words that reduce our continued use of and enjoyment of the earth. This tax would have the effect of internalizing both ecological and social costs; since all consumers would be paying the real costs, no consumers would be protected. It should also have the effect of reducing pollution.

Many of these things have been subsidized for many decades as a result of the power of special interests. The purpose of this kind of tax is to change behavior that depletes resources and discourages labor. It also can pay for the damage caused by misplaced or used resources.

Land Conversion Loss Taxes for Community. This tax would be on the conversion of complex systems to simpler, and more expensive to manage, systems, for instance, the conversion of forests to agricultural fields, or the conversion of fields to parking lots. The rate of taxation would be directly related to the cost of restoration plus the loss of productivity over the period of time before restoration.

Nonrenewables Loss Taxes for Community. Nonrenewable means functionally nonrenewable in a human lifetime, although technically such resources are renewable in the very long term as a result of very slow processes, such as continental drift and folding. Nonrenewables includes geothermal energy, as well as fossil fuels, such as oil and coal. The rate of taxation would be related to the cost of human production of those things.

Geothermal Loss Taxes for Community. Geothermal energy is the long-time flow of the energy of formation of the planet to space; it is not inexhaustible, although it may not be used up within the lifetime of the human species. It should be taxed at the community level because it is inequitably distributed, and to provide support for other energy sources that require more research and development.

Fossil Fuel Loss Taxes for Community. In a way, energy does not cycle because it is lost in equal quantities from the planetary system; however, energy may be trapped by cycles on the earth for millions of years—in coal or oil, for instance. Oil, coal, peat, and some woods are functionally nonrenewable. Geological time periods are required to produce them. Assuming 1×10^{16} grams of carbon are fixed each year by photosynthesis, averaged over a billion years, the total mass recycled of carbon alone exceeds 10^{25} grams. The total coal and oil deposits in lithosphere are estimated at 10^{19} grams. However, a small amount of detritus does not get recycled; these ordered compounds form deposits. Marine organisms produced large deposits of oil, possibly in a reducing atmosphere; forests left ranges of coal.

Oil and tar may have been first extracted near the Zagros Mountains to be used for binding tools. Since then, oil production really has meant extraction. That is the only economic means to get it. It would be expensive to synthesize. F. de Chardenedes wrote a scenario of natural technology for producing petroleum; if the amount of energy, as heat and pressure, had to be paid for at a human public utilities rate (circa 1970), the cost would be over a million dollars a gallon. If we based the cost on human labor and technology, so that 1 gallon of oil is worth a million dollars, as Buckminster Fuller also calculated, then it would be used with more circumspection.

As technology improved, wood fires also shaped metals and provided steam for power; coal fires provided electricity; oil fires provide electricity and motion. Furthermore, oil and gas can damage and destabilize ecosystems.[7] Oil use places extra carbon and pollutants in the air. A tax on oil would reduce its extraction and use, as well as encourage renewable, cheaper alternatives.

Natural Gas Loss Taxes for Community. Natural gas could be taxed at a much higher rate, a tax of 25 percent instead of 5 percent of the market value of the natural gas. This higher rate would discourage the use of natural gas, and to encourage the development of alternative and relatively renewable sources, such as solar energy, although the sun has a finite lifetime also. Natural gas reserves could be used as a transitional energy source or for special needs.

Coal Loss Taxes for Community. Coal would also be taxed at a higher rate, a 25 percent tax based on the sales price per ton, instead of a one-cent-per-ton or a dollar-a-ton or a small percentage of the sales price. This should cover, not only the cost of restoring the landscapes or mines, but also the medical costs, such as black lung, associated with the mining of coal.

Slow Renewables Loss Taxes for Community. These resources are essentially nonrenewable in terms of a human lifetime. Although some trees can mature in forty-year cycles, many require eighty, a hundred years or over three hundred. Forests themselves may take many hundreds of years to mature.

Although individual fish can mature in one to four years, fish populations require much longer time periods. Fish populations are also vulnerable with relatively low percentages as takes.

Trees & Forests Loss Taxes for Community. Taxation can be a controversial field. Many states or provinces have special private forest tax laws that exempt timber from taxation or defer the tax until harvest. Economists fight over whether the bare land is the capital and timber a 'goods-in-process,' or whether the timber is the accumulated capital. One could even state that the forest ecosystem is the capital, not just the timber. Trees, as well as other species, fresh air and water, are the interest on a forest; the whole forest itself is the capital.

In the U.S., the Oregon Legislature intended to tax timber like any other crop, while forest land would be taxed at reduced values, under an overhaul of timber taxation that cleared both houses. Constanza and Daly suggest a natural capital depletion tax could be applied to forest use that is not biologically sustainable. Sustainability here is used to mean no extrinsic costs, that is, we cannot sustain forest use at the expense of clean air and water, at the disruption of processes, and with the destruction of plant and animal communities. This would tax the use of virgin materials for commerce. Both ideas would improve forest use, but could be extended to other ecosystems.

Forest land should be taxed on its land value. Since the forest is the capital, and trees characterize the forest, the depletion tax on trees needs to be high. If enough trees are removed the forest dies. The tax should be applied so that the benefits stay local.

Fish & Fisheries Loss Taxes for Community. Wildlife has commercial value, that is, fish can be caught without being produced. The constancy of the environment, even with its changes, has allowed many kinds of animals to live in large populations. Fish have quite explicit needs for temperature, stream bed character, and cleanliness. Taxing fish takes should allow populations and habitats to recover. The community must be restored to health. This means balancing human needs with fish needs in a sustainable pattern.

Fast Renewables Loss Taxes for Community. Fast renewables are things that are usually part of daily, monthly or annual cycles, such as solar power, hydropower or wind power. They can be taxed at a relatively low rate or even a negative rate to encourage development.

Large-scale use of any of these methods of production could result in other problems. Other than the aesthetics of giant wind or solar 'farms,' concentration could cause problems with ecosystem interference and stability, as well as with human safety and aesthetics.

Wind Power Loss Taxes. Wind is a term for the movement of air, usually the result of heat differences in the air and planetary rotation. Although wind energy is not infinite, it will continue for many millions of years. The purpose of taxing it is recognize its values and limits, as well as to shift to more benign energy sources. Wind power could have a negative tax (or tax credit) of $0.02 per kilowatt-hour for electricity produced from a wind farm.

Solar Power Loss Taxes. Energy from the sun drives most cycles on earth. Even though solar energy is finite, it will continue, and increase slowly, for billions of years. Solar power could have a negative tax of 3 cents per kilowatt-hour for electricity produced from arrays of solar collectors. There would be tax credits for solar domestic or business water heating.

Hydropower Loss Taxes. The earth-moon system generates tides on both bodies. The water cycle on earth pumps water into the atmosphere and its descent can also generate power. Hydropower would have a negative tax of 1 cent per kilowatt-hour for electricity produced from large or micro generators.

Waste Loss Taxes for Community. Some waste can be considered a measure of loss from the system, that is, resources that are not captured for reuse. Many wastes come from industrial, agricultural, or mining operations and include toxic substances. A lot of waste comes from private households, in the form of paint, pesticides, and poisons. Many wastes, from solid wastes to energy, would be taxed by weight or neutralization costs and at amounts necessary to discourage their use or to encourage their incorporation into cycles.

Adjustment Taxes for Community
Adjustment taxes are regulation or correction taxes to regulate or harmonize cures and causes. They are not applied to loss but to meet the cost of recovery. For example, water returned to the watershed that does not meet purity standards would be taxed. Many forms of solid waste, such as garbage, refuse, or sludge, could be taxed by volume, at a rate of 50 cents per cubic foot. Other forms of solid waste, such as hazardous waste, tires, batteries, or nuclear by-products would be taxed sufficiently to pay for their isolation or break-down. Recyclable solid waste, such as animal excrement or rock, would not be taxed if it was recycled properly.

Sin Adjustment Taxes for Community. If people do dangerous things, and expect their government to take care of them, then government has to tax those things that result in illness. This tax would benefit people who suffer from bad habits. Due to its relatively high rate, it might reduce the consumption of addictive substances. The money from these taxes would address human consumption requiring health care, especially for health care beyond what the individual can afford.

Cigarette Adjustment Tax for Community. Inhaling the smoke from burning in plants in moderation may be stimulating, especially in ceremonial settings. However, overuse has serious health consequences. An industry exists that makes profits by selling such things to users and addicts. Like an excise tax, this tax is on sales of cigarettes, usually a fixed fee on each pack of cigarettes sold. This tax can double or even triple the retail cost of cigarettes in some U.S. states.

Alcohol Adjustment Tax for Community. Drinking the products of fermented plants may also be stimulating. It certainly has a long, rich tradition, from native peoples eating fermented berries to the large-scale production of ale in Mesopotamia after 9,000 years ago. This relatively high tax would offset the increased medical and social expenses from its use.

Drug Adjustment Tax for Community. Many communities try to control the kinds of drugs used. Many drugs, that may have greater or lesser consequences than those categorized as legal, may be illegal. Legal drugs would be taxed according to their purpose or perceived social benefit. Illegal taxes would be taxed at a much higher rate.

However, if all substances were legalized, then taxes could be used to control their use. There could be additional social benefits, such as decriminalization and deimprisonment, with savings and costs entailed.

Pollution Adjustment Taxes for Community. All forms of pollution would be taxed, especially those related to regional and global cycles of the elements necessary for metabolism of living beings. The amount of pollution that a firm or product releases into the air, water, or soil would be taxed. A pollution tax would address the 'market failures' that arise when businesses and consumers are not held responsible for the full health and environmental costs associated with their activities. Pollution taxes make polluters pay for their damages and incorporate these costs into their decisions and product prices (Specific elements of pollution, such as carbon and sulfur, are addressed elsewhere).

Water pollutants, for example, could be taxed on the basis of Biological Oxygen Demand (BOD). Dissolved oxygen is necessary to sustain fish and other aquatic life. Generally, firms that are subject to water pollution standards do not pay taxes or fees based on effluents that regulations still allow. Most of the high-volume BOD dischargers, referred to as point sources, are publicly owned treatment works, paper and pulp mills, food processors, metal producers, and chemical plants.

Industrial Adjustment Taxes for Community. Particulate matter is the general term used for a mixture of solid particles and liquid droplets found in the air. Those particles come in a wide range of sizes: Fine particles are less than 2.5 micrometers in diameter, and coarse particles are larger than that. The particles originate from many different stationary and mobile sources, as well as from natural sources. Fine particles result from the combustion of fuel in motor vehicles, power generation, and industrial facilities as well as from residential fireplaces and wood stoves. Coarse particles are generally emitted from power plants, factories and sources such as vehicles traveling on unpaved roads, materials handling, crushing and grinding operations, and windblown dust. Some particles are emitted directly from their sources such as smokestacks and cars. In other cases, sulfur dioxide (SO_2), nitrogen oxides (NO_x), and volatile organic compounds (VOCs) interact with other compounds to form particles.

A tax on coarse particles could force some electric utilities and manufacturing plants to install improved electrostatic precipitators, wet scrubbers, or other equipment to reduce their emissions and lower their tax burden. Reductions in emissions caused by the tax would be economically efficient if the additional abatement costs were less than the social benefits from reduced pollution.

Opponents of this kind of tax argue that it would impose an excessive burden on firms that already incur costs to comply with current standards. Firms have escaped their justified burdens of safety and efficiency, so this is not a good reason. To the extent that the tax would raise the price of energy generated in this way, it might be regressive. But, once the price of energy reflects its true costs and effects, and once many unfair taxes are removed, then some disadvantaged people and businesses can afford to pay a larger percentage for energy costs.

Agricultural Adjustment Taxes. In order to make its impressive gains, agriculture tries to control more of the growing process of crops; it does this by adding fertilizers and pesticides, as well as by using special, energy-intensive machinery. These things would be taxed to offset their costs and effects.

Pesticides Adjustment Tax. Pesticides can have a number of adverse impacts on human health and the environment, imposing substantial costs on society. These include direct financial costs, such as the treatment of water, and wider environmental costs, such as loss of biodiversity, which are much harder to value.

Pesticide impact reduction is a prime candidate for the use of economic instruments. A pesticide tax would greatly benefit farmland birds. The tax could invoke the 'Polluter Pays' principle and reduce the inappropriate use of pesticides, as well as raise revenue to pay for solutions to the environmental impacts of pesticides. Preliminary economic analysis shows that a well designed tax would not significantly damage farming incomes or competitiveness.

Fertilizer Adjustment Tax. Fertilizer is used to increase the productivity of crops, although rarely is that productivity equal to the that of the original land cover. Fertilizers rob plants of root mass and foods of nutritional values. This means that more of the crops have to be planted, and people that have to eat more of the plants to get the same nutrition.

The application of environmental tax shifting to pesticide and fertilizer use could provide an incentive to minimize pesticide and chemical fertilizer use, and to generate revenue to allow for tax relief in other areas.

Water & Other Adjustment Taxes. To encourage crops to grow in areas that have low rainfall, such as central Washington state in the U.S., farmers irrigate their crops, even crops that are not native to the area, with water drawn from an aquifer. Since water from aquifers is often renewed at very low rates, it would be taxed at a rate designed to reduce its use down closer to replacement rates.

Personal Pollution Adjustment Taxes for Community. A tax would be charged on personal pollution, from solid waste to discharges from machines. A fuel tax would be imposed on the sale of fuel to cover the calculated pollution from its use. In the US, the funds are often dedicated to transportation or even roads, so that the fuel tax is considered by many to be a user fee. In other countries, the fuel tax is a source of general revenue. Sometimes the fuel tax is not imposed on fuel that is not intended for transportation, such as fuel used to power agricultural vehicles or for home heating oil. This creates an economic incentive for the illegal use of fuel. The solution would be to tax all fuel regardless of use.

Because of the inelastic nature of the demand for fuel, in the short run the tax would be an effective source of revenue. In the long run, however, people would adjust their consumption; over a period of years, people would consume less as the price increases, by driving less or by buying more fuel-efficient cars or furnaces. A fuel tax could be a way to reduce reliance on environment-damaging fossil fuels.

Sale of Heritage Items Adjustment Taxes. A heritage tax would be applied to any thing considered part of the natural or cultural heritage, such as special buildings or landscapes. It would also include unique art works produced by the culture. It would include a tax on the export of such things without a value-added component, such a raw logs or unprocessed resources.

Financial Speculation Adjustment Taxes.
The speed and detachment of money may artificially change the values of things that should remain uninflated or undepressed. A speculation tax, like a Tobin tax, is a small tax on each international financial transaction.

There are many potential benefits from a modest tax on financial transactions, such as the buying and selling of shares of stock or blocks of foreign currencies. Such a tax would have the effect of reducing short-term speculation in these markets, thereby making them somewhat less volatile. It would slow capital movements. It would also cut back some of the economic resources that are wasted in these transactions, since if the number of trades declines, the money spent on these trades would decline as well. In addition, it would make the tax code fairer, since most financial speculation is conducted either directly or indirectly by wealthy people. Just as poor and moderate income people pay taxes when they gamble at a casino or buy a state lottery ticket, a speculation tax would simply be applying a comparable tax to gambling in financial markets. Finally, a speculation tax could raise an enormous amount of revenue for a community or state.

Distribution Taxes for Community
Distribution taxes are the same as reapportionment for luxuries and large incomes. Many large incomes are so large that they are heroic. Combined with heroic inheritance and profits, these incomes essentially remove their receivers from any sense of local community by concentrating wealth.

Heroic Possessions Distribution Taxes for Community. A heroic possessions tax is a tax on products that are not considered essential, in other words, luxuries. A luxury tax is similar to a sales tax or VAT, except that it mainly affects the wealthy because the wealthy are the most likely to spend heavily on luxuries such as expensive cars, boats, houses, or jewelry. One might even consider this tax as a 'nonuse' tax, since many luxuries are not often used and are taken out of circulation for others.

Although a luxury tax of some sort may be justified on many things, the regulations would have clarify implementation in order to achieve at least some degree of equity. Although the tax would be collected at the community or state level, it might be coordinated at the national level, considering the multiple residences of those who can afford other luxuries. Regulations have to broaden the definition of luxury. It is hard to argue that a luxury tax is unjust and counter-productive, considering the massive suffering and starvation outside the gated compounds.

Heroic Income Distribution Taxes for Community. A heroic income tax would be applied to incomes over a certain amount, as determined by a ratio of one to ten. Herman Daly and John Cobb quote a range of the acceptable inequality of income at ten to one, although some corporations, like Ben & Jerry's ice cream, used to limit it to a one to seven ratio. As Cobb and Daly point out, the idea of unlimited inequality works against the notion of community. As they also wisely point out the goal of a community is not some perfect ideal of equality, but a limited inequality that allows individual differences to show, and individual rewards for luck or skill, but also allows others to catch up. And, this idea, rather than being new, is explicit in many biblical accounts of Hebrew laws governing landholding and usury, among other things. So, this tax rate would be quite high for those with heroic incomes. Paying such a tax would make these people heroes and in fact they might acquire heroic status for paying.

Heroic Transfer Distribution Taxes for Community. A transfer tax is a tax on any transfer of wealth, including inheritance, death duty, estate, or bequest. It has similarities to death taxes, inheritance taxes, and estate taxes, but it is much broader. It is applied to any transfer of wealth, to spouses, children, relatives, and even bequests to trusts, foundations and charities, and not on the total wealth. This would include payment of life insurance benefits or financial account sums to beneficiaries.

This tax would require relatively heavy regulation. This tax would serve to prevent the perpetuation of wealth, free of tax, within wealthy families. It would work towards reducing the accumulation between generations, thus leveling the field more for the next generation.

Heroic Profit Distribution Taxes for Community. A community has to be responsible to keep the economic playing field level, to use a popular sports metaphor. In the event that a person or corporation records a profit that could distort the community, then the community has to limit that profit through taxes to not more than ten times the mean profit in the community (not the average or median).

Discussion of Transitional & Missing Taxes for Community
Under this new tax scheme, there would be no taxes on buildings, equipment or inventories. There would be no corporate income tax, although, as Daly and Cobb suggest profits would have to be distributed to all shareholders as income. There would be no personal income tax, no property tax, and no sales tax. Why? Because things people want or work for should be encouraged. Once the real price of oil and other goods has settled out, no other taxes are needed for value-added things.

In order to reach this position, however, there may be a series of transitional taxes, such as an added tax on all new vehicles.

License Privileges for Community
A license is a formal authorization by law to do something, such as to marry, hunt, or practice law or medicine.

Marriage License for Community. A marriage license is permission from a legal authority for the marriage of two people to be performed. The requirements differ depending on the time and place: Licenses to marry have been granted since the Middle Ages. Valid marriages can occur without a license, for example, by obtaining pardon for having married without license, or by cohabitation and representation as husband and wife in jurisdictions permitting common law marriage.

Every state in the United States issues marriage licenses. After the marriage ceremony, both spouses and the official sign the marriage license; some communities or states also require a witness. The official or couple then files for a certified copy of the marriage license and a marriage certificate with the government.

What is the purpose of such a license? Health? Tracking? People who wish to live together with the formal approval of a government need to meet certain requirements of residence or health. Some groups believe that needing to obtain a marriage license from the State in order to be married is unnecessary or immoral. Combined with other rights or privileges, such as having children, a marriage license would be required to indicate preparedness and intent, as well as to collect benefits.

Reproduction License for Community. A reproduction license would be permission from a legal authority for two people to have offspring. The requirements would differ depending on the time and place. The purposes of such a license would be several, starting with qualifying the parents, through a reading program or knowledge of health issues, such as smoking or drinking, and continuing with the health and safety of children, and with consideration of the effects on the health and fitness of the entire gene pool and on society.

Parenting skills would be tested and improved, much as certification and licensing improves day-care and educational providers. Poor parenting can be linked to juvenile delinquency, crime, and sexual abuse. Children with poor personal and social abilities can burden the social system, especially the educational system; as adults they can continue to burden the system and produce another maladapted generation.

The penalty for not having the license might range from a fine to reallocation of the child to another family or institution, although that alternative would have to have numerous safeguards for the health and safety of the child.

One incentive for having the license might be a monetary rebate. Another incentive could be an extra coupon towards education or luxuries. This license would not limit the number of children or help with the planning of institutions, such as schools—that would be the purpose of tradable vouchers to have children.

Driving & Vehicles Licenses for Community. A driver's license is an official document which states that a person has the qualifications to operate a motorized vehicle, such as a motorcycle, car, truck, camper, trailer, or a bus. Driver's licenses are generally issued after the recipient has passed a successful driving test and proven that they meet the age, education and understanding requirements. Different categories of licenses may exist for different types of motor vehicles. The difficulty of the driving test may vary considerably between jurisdictions, but perhaps be above minimum national standards.

Vehicles themselves must also be licensed for use and safety. The term 'motor vehicle' includes automobiles, trucks, buses, campers, trailers, and motorcycles. Motor vehicles are valued by year, make, and model in accordance with a formal vehicle valuation manual. Values are based on the retail level of trade for property tax purposes.

Voting License for Community. A license to vote would be required to demonstrate that the voter is living, lives in the community, speaks the language, and understands the basic issues. This would prevent some kinds of fraud.

Wavelength License for Community. Any use, for broadcast or communication, of common frequencies of the spectrum of wavelengths, would be licensed by the agency responsible for regulating interstate and international communications by radio, television, wire, satellite, or cable.

Business License for Community. Virtually every U.S. state licenses businesses. A business is defined as any commercial enterprise, trade, occupation, profession, or activity engaged in, conducted, carried on, advertised, or held out to the public to be a business by any person, agent, or employee for the purpose of gain, benefit or advantage, either direct or indirect, with the principal objective of livelihood or profit through repetitive means.

The business license certificate is evidence of having met community requirements and that a fee has been paid for doing business within the limits of a community or state. Most codes requires that people obtain a license when conducting any business activity within the community, even if the business is located outside the area and has a license from another area. Other permits may be required to operate a business.

The fee imposed is solely for the purpose of obtaining general revenue. Business fees could help pay for community services like roads, fire, police and other community services that benefit businesses, business owners and the general public.

Collecting License for Community. Anyone who collected plants, animals, fish, or any living substance for its body, compounds or DNA, would be required to have a license. The license would certify that they have the knowledge to collect, and it would limit the number of people collecting. Issuing licenses would help to protect commons, such as parks and wilderness areas, as well as to coordinate the timing of collection and the limits of collecting.

Weapons License for Community. A weapons license would verify that knowledge and experience requirements are met, and reviews and researches of criminal history have been conducted for information that might preclude the issuance of a license, such as a criminal record or history of violence.

The community could create a bureau responsible for the issuance and denial of licenses. This bureau would receive and examine licensing applications for statutory compliance and verify the applicant's eligibility for licensing through former employers, educational facilities and examination of criminal history records. The kinds of weapons could be limited to knives, arrows, single-shot firearms and multi-shot fire arms; certain weapons, such as automatic guns, tanks, bazookas, and missiles, would not be licensable.

Applying for a license to carry a weapon for self-defense is a right of the law-abiding members of many communities in many nations. Weapons can be carried responsibly, properly and safely.

Fees & Tolls for Community
Fees could be charged by the community for use of community space or renewal cycles, for example, visiting fees or sanitation fees. Tolls could be charged by the community to cover extra expenses for transportation, for instance, bridge tolls or road tolls. Special fees could be charged for tourist attractions and for heritage sites.

Labor Use for Community
Many forms of volunteer work take place in a private setting or for an NGO dealing with social problems or environmental conservation. People volunteer for hospitals or libraries, food programs or the humane society. Many volunteers are coordinated by large networks for specific services. A community could encourage this or support it, with many kinds of recognition and reward.

Payout for the Community

The community has to pay for its existence, as well as the expenses, amenities, and services that it trades for its existence. The payout, or outgo, is available from income, and should balance income, although savings accounts and debt-load can expand or contract temporarily.

In economics, business and accounting, a cost is the value of inputs that have been used up to produce something, and hence are not available for use anymore. In business, the cost may be one of acquisition, in which case the amount of money expended to acquire it is counted as cost. In this case, money is the input that is gone in order to acquire the thing. This acquisition cost may be the sum of the cost of production as incurred by the original producer, and further costs of transaction as incurred by the buyer over and above the price paid to the producer. Usually, the price also includes a mark-up for profit over the cost of production. Costs are often further described based on their timing or their applicability.

When a transaction takes place, it typically involves both private costs and external costs. Its private costs are the costs that the buyer of a good or service pays the seller. Its external costs, which are also called externalities, in contrast, are the costs that people other than the buyer are forced to pay as a result of the transaction. The bearers of such costs can be either particular individuals or society at large. External costs are often nonmonetary, and are difficult to quantify for comparison with monetary values. They include things like pollution, things that society will likely have to pay for at some time, but are not included in transaction prices.

Social costs are the sum of private costs and external costs, that is, both the costs internal to the firm's production and external costs not included in the firm's production. For example, the purchase price of a car reflects the private cost experienced by the manufacturer. The air pollution created in the production of the car is an external cost. The manufacturer does not pay for these costs and does not include them in the price of the car, so they are external to the market pricing mechanism. The air pollution from driving the car is also an externality. The driver does not pay for the environmental damage caused by using the car. A psychic cost is a subset of social costs, for example from endangering pedestrians, that specifically represent the costs of added stresses or losses to the quality of life. A community has to consider all costs for as long as possible into the future.

Issue vouchers for Community
A voucher is an economic warrant and guarantee for goods and services. A voucher is a certificate which is worth a certain monetary value and which may only be spent for specific reasons or on specific goods. Examples include, but are not limited to, education, children, and medicine. The community would issue certain vouchers for all citizens. Vouchers would be an effective way of dispersing limited resources and reducing inequity.

Basic Income Vouchers for Community. A basic income voucher would be issued every month for community residents. This voucher would cover minimum clothing, housing, and food.

Medical Vouchers for Community. A basic voucher could be issued by the community to allow the consumer to choose basic medical care. Extra care could be acquired through insurance companies that use their own voucher system. In some respects vouchers resemble credit cards, although they would be more specialized. Vouchers may vary in value over time, and may be specialized for use in the separate phases of medical care.

Productivity Resource Vouchers for Community. Within a calculated optimum (see elsewhere), every community would receive a number of vouchers for its population for the optimum or maximum use of resources and processes. Each individual would receive a set of vouchers for specific resources, such as productivity or carbon. Each community could make decisions on population size and wealth depending on other values or trade-offs.

Carbon & Other Element Vouchers for Community. Every community could be issued a number of vouchers equal to the carbon-bearing products of it population and their standard of living, based on a world standard. These would determine how much was used in each national system, regardless of its population or the intensity of its exploitation. Each community could be given vouchers for each element or compound, such as phosphorus or iron. These would be distributed among the population, according to a formula.

Vouchers for Water, Air & Other Compounds. Everyone has basic requirements for breathing, drinking, or getting materials. The requirements of living would not require vouchers, except under extreme circumstances of overpopulation or overconsumption. Trying to issue vouchers for these things, on a personal level, might be too expensive and too invasive.

Education Vouchers for Community. An education voucher, commonly called a school voucher, is a certificate by which parents are given the ability to pay for the education of their children at a school of their choice, rather than the closest public school to which they were assigned. This kind of voucher could be accepted at a publicly supported school, although it might not pay for a complete private education. The school itself could decide to accept or not accept a student voucher, depending on qualifications and load limits.

Child Vouchers for Community. Several years ago, Kenneth Boulding recommended issuing women a marketable license permitting them to have a limited number of children, saying that the right to have children should be a marketable commodity, bought and traded by individuals, but limited by the state. For instance, every person would be issued a 0.5 replacement voucher share at birth. This number would change every year depending on population and support. For instance, for 2007-2008 the value might be 0 (or 0.1); for 2019-2020 it might be 0.3, and for 2034-2035, it might be 0.55.

These vouchers would be tradable and combinable. These vouchers could be traded so that, when a couple had a full voucher (1.0 or over) they could have a child. The voucher could not be used until the holder reached a majority age. This would result in a slightly declining population, even with a small percentage of cheating. The value could be adjusted once the population was at the desired target, perhaps at an optimum cultural carrying capacity.

Others Kinds of Vouchers for Community. What other kinds of things could be distributed using vouchers? Should there be vouchers for luxury? In a random lotto system for fairness? Or should it be a reward for generosity or some other virtue?

Community Costs for Health & Maintenance
The community has to take care of itself and to maintain itself as a separate entity. The community has to be healthy and vibrant. It has to build and maintain the infrastructure used by its citizens, from heroic architecture and dramatic boulevards to parks, theaters, and public areas. It has to pay the costs of these necessities.

Promote Self-reliance of Community
Self-reliance is the basis of security for a community. It is indication that the community has adjusted to its environment. And, that it is strong enough to resist some of the blandishments of other cultures, as well as recover from environmental disturbances. Communities can be self-reliant by producing enough food and shelter, by limiting their population to what can be produced, by using local products and raw materials, from soil and minerals to plants, by using general and not specialized machines, by having multipurpose factories, by networking with other communities, and by doing without things that are not needed, such as food additives, or aluminum bottles.

Operate Community Government
A community has to operate its local government. Leaders and representatives have to be elected and paid. It has to determine and fund the level of local infrastructure, from roads to schools. It has to establish and fund special departments for joblessness or old-age difficulties. It has to control communications and trade exchanges with other communities.

Encourage New Business
The community has to encourage the formation of new businesses and provide ways to extend the useful lives of businesses beyond the first year. The community can attract entrepreneurs, then set up programs to develop new businesses through incubators or special economic programs. New businesses often contribute disproportionately to innovation and efficiency.

Assess Community Performance
Adam Smith developed the argument that the only fundamental way of assessing the wealth of a nation is by examining the manner in which it uses its labor. This is true in a very basic way, since food, clothing and housing are products of labor. And, those countries with the highest standards have the largest output per person-hour. However, his thesis is not complete. It attributes no inherent value to the raw materials upon which agriculture and industry are based; these are treated as free gifts of nature, available in infinite amounts. Labor is an incomplete means by which the total economic well-being of nations could be assessed.

The assessment of personal or cultural wealth, for instance, is mostly psychological; wealth may be measured by how many valuables one has, which may be physical, like feathers, cattle, gold, or land, or by how by much status one enjoys, which may be behavioral, such as enjoying deference or a good reputation. Every assessment has degrees of subjectivity. A community has to relate its goals and images to its values and products, as well as to the satisfaction and happiness of its citizens.

Evaluating Leadership
Making these changes occur all at once would seem unrealistically disruptive, but would it be? Would it be more disruptive than an earthquake or a war, than runaway inflation or economic collapse? These things could all be done if people were convinced by their respected leaders that this was the best way to proceed. Given the fact that the characteristics that make people lust for leadership often detracts from their abilities to lead, perhaps we need a form of draft to get leaders, or perhaps we need to consider persons who have been groomed for decades, like the Dalai Lama.

We need to examine different models of power and authority: Autocrat, guru, revolutionary, or transformer, as W.I. Thompson suggests. The transformer (or Elder) has authority but no power; the autocrat has both; the revolutionary has power but no authority; and the guru has neither. Perhaps we should consider the guru as leader, to transform the world using the metaphor of water to follow the flow of things.

Until we can do things naturally and spontaneously, we must try to act our way into right thinking, while thinking our way into right action. It is a form of intelligence — understanding principles and trends of natural affairs so that the least energy is used dealing with them. This is also the unconscious intelligence of a whole organism and the whole community.

Contributing at the National Level

Individual and community efforts may be most important, but action at the national level can coordinate changes or make them a matter of policy. Making the policy can feed back positively into community and individual actions as well. The national level itself is quite fuzzy. Although people group into families and communities, there are many possible groupings between the individual and the national. The responsibilities of each group may be determined at the national level, or certainly through a process involving the national, regional, local and personal levels. Thus, some things may be more appropriately controlled at a county, state or province level, such as collecting taxes for roads. Other things, such as health insurance, may be best directed at a national level, to ensure the equity of distribution.

Catastrophic & Normal Steps for a Nation
Leadership at the national level could balance the dramatic changes suggested here—perhaps one leader could act like a shaman in Desana society, who is responsible for acknowledging the ecological limits and suggesting the behavior required to stay within those limits. Some of the following steps may seem revolutionary. Unfortunately, revolutions have the connotations of violence and overthrow. Revolutions can be as quiet and regular, and unthreatening, as the turnover of an axle on a wagon or car. Thomas Jefferson suggested that little revolutions, every couple of decades in the U.S., could make the experiment fresh, as well as break up unproductive hierarchies of power. We could start these little revolutions with many small steps, as long as they did not contradict cultural norms.

The first revolution has to be to adopt a new attitude. Adopting a catastrophic psychology for the nation, to address current and imminent losses, is less revolutionary and a more appropriate response to larger scale catastrophes, such as the national loses of biological and cultural heritages. Poverty and inequity are growing problems in every nation. These are reasons to adopt an extreme attitude towards survival.

Changing Nation to a Larger Self. As humanity is extended throughout cultures and nature, the human self interpenetrates so much that it becomes a larger self. The nation is at a level of the stereotype of the people of a culture, the sum of their behaviors, as it were. It is the larger self within the enveloping self of the environment. The nation has to change itself, and to recognize the extension of itself in nature, before it can make other changes. National behavior emerges from the individual behaviors.

Define & Secure Borders for a Nation. Borders are a crucial part of a nation, to keep its identity and define its territory. Borders form the skin that acts as an organ to keep vital processes and cycles inside and allow needed energy and materials to come inside. A nation has to decide what kind of borders to maintain, with what degree of openness. The amount of openness allows the renewal and protection of some areas. Leopold Kohr points out that much violence between nations is the result of bad divisions between them. Two alternatives to bad division are unification or good division.

Balance Isolation & Connectivity for a Nation. A nation has to control its relative degree of isolation from other nations. Isolation offers many advantages and a few disadvantages. It allows differentiation, of language and invention, that can reinforce the identity of a culture. It can offer safety from conquest by overwhelming force. Too much isolation, however, can lead to lack of stimulation and to ingrown customs.

A nation has to regulate the amount and kinds of trade. It has to decide how to participate with other nations, to ignore them or to compete with them, and how. It has to play a part in regional and global processes.

Balance Immigration with Emigration for a Nation. Some people will always be uncomfortable with the dominant culture of a nation and wish to leave it. Others will always be attracted by the unique qualities of another culture and want to join it. Nations may also try to balance certain skills and subpopulations with its overall goals, depending on their trade specializations.

Secure Treaties with Neighbors for a Nation. In addition to having secure borders, a nation has to normalize relationships with its close or regional neighbors. Such official agreements would spell out the extents of exchanges.

Set Standards for a Nation. Standards are models or examples of quality or value established by authority or consent that can be repeated as procedures. Standards can be used to certify practices for harvesting or manufacturing. A nation has to set standards for requirements, tariffs, protections, and for emissions and other things. A nation should also set the official standards for those things necessary for political interactions or economic production.

Ecosystem Standards for a Nation. Ecosystem standards would start with zoning. In a typical meso-ecosystem, a minimum area, comprising at least fifty percent, should be left wild. A smaller area, perhaps thirty percent, should be set aside for conservation of forests and rangelands. Based on the planned population, at least fifteen percent should be set aside for agricultural areas to feed the population. City and artificial areas should be restricted to the remaining small percentages, of the least productivity. In cases where the city area exceeded that maximum percentage, an equal area of rooftops or pavement areas would have to be dedicated to agricultural activities. Thus, there would be no limit to house density, only coverage area, and that would be directly related to natural primary productive areas.

Social Standards for a Nation. A nation has to have standards for the health of its people. It has to describe the minimum kinds of housing, utilities, and cleanliness that are acceptable. It also has to decide how much looseness and cheating are allowed, and how much punishment to apply to those who are caught.

Economic Trade Standards for a Nation. Tariffs may be needed for the protection of the manufacturing of special trade items. Nations should have the right to reject trade items that go against their laws or moral codes, or that may be hazardous to people or their environment.

Technology Standards for a Nation. Standards need to be set for those things related to transportation, such as the extent and quality of roads, as well as the safety and efficiency of vehicles. These standards might try to reverse trends towards large-size, inefficient trucks and sports utility vehicles (SUVs); the average SUV emits two to three times as much greenhouse gas as the average compact car. Even if a van is necessary to transport the whole family, there are ranges of choices in terms of quality and gas-mileage. The choice can make a ton of difference in CO_2 emissions every year, literally. Gasoline-powered vans and cars can be converted to hydrogen or waste oils.

Unlike those for aviation, most countries do not have a feedback process to make automobiles or boats safer. Every nation could create a department for analyzing every accident and making requirements for standards and behaviors and infrastructure changes.

Many roads could be decommissioned after being inventoried and planned for optimum coverage. Many roads in forests and wilderness areas need to be deconstructed for the health of those systems.

Few nations have national building standards, either. Standards for buildings, especially for air-conditioning, need to consider efficient alternative forms of power, including solar, wind, and geothermal (and tidal, current and heat differential systems in coastal areas).

Establishing Rights for a Nation

Rights, as an extension of ethics, are simply rules for living together. A nation has to codify those rights so that they will respected by all the residents, not just those who share the dominant culture.

Rights for Nature in a Nation

Rights seem to follow the expansion of the sphere of ethics, as formal statements of intuitive knowledge. Thus, rights for nature are being considered by many people. Paul Shepard says the argument is not new, and that its application is ambiguous because 'unlimited rights' will conflict with human interest. But, there are two bad assumptions: That human interests are not ambiguous — they are — and that animals will be granted unlimited rights — they will not.

The strongest argument for rights is interrelatedness in communities, which is the basis for assigning rights to nature. Garret Hardin considers interrelatedness, but interprets it narrowly. He considers rights as rules of competition; every right is a ploy in the struggle for existence, and every right implies an obligation to furnish it. This is good as far as it goes. However, life is more than competition; it involves cooperation and play. Rights are formal rules for living together. It would be foolish not to assign rights to animals, plants, and the earth because of contractual formalities.

Humanity has taken its own opportunities. These opportunities have been codified for centuries as rights. Now, we must allow other beings equal opportunities. The interrelatedness of life dictates the interrelatedness of rights. And these rights are necessary to the integrity of the whole planet. Humanity developed in a community of animals and plants. The quality of human life has always depended on the quality of animal life. Animals have sensations and feelings, as important to them as ours are to us.

Furthermore, the extension of rights to animals and plants does not deny any traditional human rights. Animals should be accorded higher moral regard and legal standing to reflect the intrinsic worth afforded by their existence and sentience. Welfare laws to conserve species and to guarantee humane treatment in research, transportation, and slaughter indicate a growing concern among people. A new ethic can keep animals free from human intervention, prejudice, or overuse. The intrinsic worth of animals is independent of instrumental values imposed on them by us.

One problem with the current legal system is that all nonhuman beings are given the status of inferior human beings, legal incompetents, thus keeping humans in a guardian role. A new legal category is needed that would respect the existence, competence, and excellence of natural beings. Christopher Stone recognizes that the judicial system has granted rights to a variety of inanimate holders, trusts, corporations, and nations, for instance. The legal system already operates with fictions, so the extension to natural entities should not present an insurmountable problem.

Space to Exist & Opportunity to Flourish. Every species has to be allowed the opportunity to live, even species that we fear or dislike, such as sharks or viruses. We do not know how these species contribute to the whole process of nature. Giving other species opportunities does not mean sacrificing any human needs, just limiting human influence and interference.

In our control of conservation or artificial areas, which include many wild species, we can imitate the process of ecosystems by allowing birds, bats, and other animals opportunity to distribute seeds and energy to other areas or to access their prey, which may be our 'pests.'

Freedom from Premature Death Extinction & Suffering. Animals do not need to be saved from natural death, which is a great regulator of life, but from unnecessary suffering, experimentation, and premature extinction. The world would not be a better place without sharks, silverfish, rats, cockroaches, or hyenas. They need their own places, where they can take their own opportunities, live or die. The places, entire ecosystems, need to be saved. If we diminish variety in nature, we debase its stability and wholeness, which we need to survive.

Establish Rights for People in a Nation
Humanity has the right to coexist in healthy diverse, stable conditions. The basic things expected by people of their political and economic systems are simple. They are: Equality of opportunity for anyone; jobs for those who can work; security for those who need it; the ending of special privilege for the few; the preservation of civil liberties for all; and, the enjoyment of the fruits of scientific progress in a constantly rising standard of living.

Human rights refers to the concept of human beings as having universal rights, or status, regardless of legal jurisdiction or other localizing factors, such as ethnicity and nationality. The existence, validity and the content of human rights continue to be the subject to debate in philosophy and political science. Legally, human rights are defined in international law and covenants, and further, in the domestic laws of many states. However, for many people the doctrine of human rights goes beyond law and forms a fundamental moral basis for regulating the contemporary political order.

Human rights are taken to be inherent, universal, indivisible and inalienable. This means that everyone has them, they are the same for everyone, all are equally important, and they can not be taken away. Although a right can not be taken away, it can be violated.

Human rights can be divided into seven categories: Civil Rights (equality before the law), Political Rights (participation in government, life, liberty), Economic Rights (right to work for a living wage), Social Rights (having children, education, health care), Cultural Rights (preserve a cultural identity, language, practices), Environmental Rights (a healthy environment, clean drinking water, unpolluted air), and Developmental Rights (the rights of nations to control their own resources).

Civil and political rights are sometimes divided into negative and positive rights. Negative rights, which follow mainly from the Anglo-American legal tradition, denote actions that a government should not take. These include right to life and security of person; freedom from slavery; equality before the law and due process under the rule of law; freedom of movement; and freedoms of speech, religion and assembly. Positive rights follow mainly from the Continental European legal tradition, which denotes rights that the state is obliged to protect and provide. Examples include: The rights to education, to a livelihood, and to legal equality. Positive rights have been codified in the Universal Declaration of Human Rights and in many twentieth century national constitutions.

Human rights can also be based on the 'natural' moral order described by religious precepts. Religious societies justify human rights through religious arguments. For example, liberal movements within Islam have tried to use the Qur'an to support human rights in a Muslim context.

Basic Human Rights in a Nation
Nations have to decide the kinds of rights and guarantee them to their citizens. Some of these basic rights should always include the right to a healthy environment or the right to be secure.

Right to Healthy Environment Air Water. Principle 1 of the Rio Declaration of the UN states that human beings are "at the centre of concerns for sustainable development. They are entitled to a healthy and productive life in harmony with nature." While this statement fell short of recognizing a healthy environment as a basic human right, it points in that direction. People need to be assured of having clean air, clean water, and healthy ecosystems.

Right to be Secure. The right to be secure can mean many things. It can mean being free from invasion of your home. It can mean access to wilderness or land where you can provide for yourself. Insecurity is a way of life for people living in poverty; it means that they might survive, if things improve and if nothing goes wrong, but neither scenario is likely without eutopian change.

Right to Opportunity for a Minimum Home. If you cannot afford a home, you can ask for help. If you cannot get help, you can build your own. If you cannot build one, you can live in a group that shares one. The right is in the opportunity, not in the ownership. Most people can make or buy homes, if the circumstances are promising.

Right to Opportunity to Minimum Work. Every person should have the right to work and to receive a living wage for their work. Some nations are unwilling to support nonworking adults and deny them government assistance. Other people who do sacrifice and work very hard, may not earn enough to lift themselves and their children out of poverty. When current economic and legal arrangements hurt individuals, families, and communities, then something needs to change. A change in the laws, or in the Constitution, could provide every citizen with the right to an opportunity to work for a living wage.

A nation must guarantee everyone an opportunity to work at a living wage. Most people want work, meaningful work. People want to contribute to their own well-being, as well as to that of their family, community and nation. It is in the common interest of the community and nation that people who work should not be poor or dependent on others for support.

Millions of people are seeking work. Organizations, such as Oxfam, try to help people and communities to achieve 'sustainable livelihoods,' that is, a means of living that can maintain itself over time, and can cope with and recover from shocks. Oxfam has found that even small shocks, such as a broken washing machine or an unexpectedly large fuel bill, can trigger an imbalance of income and expenditure, and can lead to a downward spiral into debt and despair. There are many factors, such as lack of access to affordable credit, which force people to borrow from moneylenders at extortionary interest rates. However, there are many factors, such as social networks, education, personal skills, and accessible transportation, that contribute to the sustainability of livelihood.

Tradable Rights
Many rights, such as the right to reproduce, are partial social rights. As such they must be traded, through a voucher system, to acquire the right to an entire child or for a certain level of luxury. Other rights, such as the right to a healthy environment, can be traded for luxuries or larger populations.

Right to Reproduce. People have the right to have children, but this right changes in the context of overpopulation, as well as child mistreatment and abandonment. The right is limited when natural or social services are limited.

Right to Share in Luxury. What is luxury? How can we deal with luxury goods provided by a free market intent on maximizing profits? Luxury, a measure of those things wanted to improve happiness, is being hoarded by the rich or by the middle classes in rich nations. Runaway growth is exacerbating inequalities. If the trends continue without change — not redistributing income from higher to lower-income consumers, not shifting from polluting to cleaner goods and production technologies, not promoting goods that empower poor producers, or not shifting priority from conspicuous consumption to meeting basic needs — then the problems will get worse.

An equally critical issue is not gross consumption itself, but its patterns and effects. Inequalities in consumption are stark. Globally, the 20% of the people in the highest-income countries account for 86% of total private consumption expenditures — the poorest 20% of people get a minuscule 1.3%. More specifically, the richest fifth: Consume 45% of all meat and fish (the poorest fifth only 5%); consume 58% of total energy (the poorest fifth less than 4%); have 74% of all telephone lines (the poorest fifth 1.5%); consume 84% of all paper (the poorest fifth 1.1%); and, own 87% of the world's vehicle fleet (the poorest fifth less than 1%).

And, consider the following expenditures that reflect the priorities of wealthy nations: $8 billion for cosmetics in the United States; $11 billion for ice cream in Europe; $12 billion for perfumes in Europe and the United States; $17 billion for pet foods in Europe and the United States; $35 billion for business entertainment in Japan; $50 billion for cigarettes in Europe; $105 billion for alcoholic drinks in Europe; $400 billion for narcotic drugs in the world; and, $700 billion dollars for military spending in the world.

Compare that to what was estimated as additional costs to achieve universal access to basic social services in all developing countries: $6 billion for basic education for all; $9 billion for water and sanitation for all; $12 billion for reproductive health for all women; and, $13 billion for basic health and nutrition for all.[8] These numbers enforce the stark inequalities in consumption.

What would happen if everyone tried to exercise their right to some luxury? Would those spoiled by, and now needing, high levels of luxury refuse to share it?

Other Rights
Other rights, such as for basic ecological services, could be controlled by issuing tradable shares for those services. The number of shares would depend on calculations for resources and optimum populations.

Work to Establish Justice. Although many new rights are being extended to cover all of humanity and much of nature, new nations can demand to participate in a common justice. Rights and their obligations, after being reduced to principles of equity, can be addressed by justice, through standards and laws.

A principle of justice based on need can be extended to the ultrahuman community. To be sure, it needs to be altered to account for unconscious, interdependent beings.

Represent All People of Nation. A nation has to give equal representation to everyone in its borders. Although distinctions will be made between citizens, noncitizens, visitors, and tourists, each will be represented to some degree.

Obligation to Protect Rights & Privileges. A nation is obligated to protect the rights and privileges that have been defined and agreed upon. Citizens will have more rights and privileges than noncitizens or visitors, because they pay for them with taxes and participation.

Obligation to Meet Basic Needs of People of Nation. A nation is obligated to provide for the basic needs of its people, and, for this reason, has some decision in how many people there are and how they can sustain themselves.

Integrate Everyone into the Society of a Nation. A nation needs to be able to integrate people into its society. This often happens with a dominant culture and language, but it has to happen if there are equal cultures or several minority cultures. The nation has be decide how people can communicate, and how they can minimize conflict or stress.

Terrorism has become a growing problem. Terrorist acts have killed thousands of people in many nations, usually for political or religious reasons. It is very difficult to anticipate and eliminate terrorist threats, especially random suicide bombings; winning a war on terror is probably not possible. However, there are some actions that may gradually reduce the need for terrorism. First, have all people show respect for the religions and cultures of others; reduce or censor the number of stupid burnings, writings and pronouncements. Next, and this will be difficult and slow, try to correct the vast and painful historical inequities between nations, classes and individuals; this would involve many simultaneous actions, such as paying living wages, taxing high incomes at high rates; drastically reducing military spending; and investing in all the public infrastructure, from jobs and food to parks and libraries (in every nation). The GU could work to arrange resolutions to long-standing ethnic or political feuds that waste so much time and life. Finally have nations and groups help one another, starting with international volunteer programs run through the GU and continuing through organizations such as a universal Peace Corps or Red Cross. Once people communicate, many stereotypes and hatreds may evaporate. But even then, some people will always be unhappy and have access to weapons.

Amend Constitution to Control Corporations & Groups. Every nation creates a formal or informal constitution. The Constitution guarantees a series of rights, such as protection, ownership of property, and a healthy environment. It also guarantees a series of freedoms, such as expression, religion, choice, assembly, association, and the petition of grievances. It prescribes some duties, such as voting or respect for others. And, it has a number of goals: Stability of government through transitions; predictability of the laws; and, common rules shared by all. In many nations, it may be necessary to amend the constitution, not just to reflect new international standards and agreements, but to reflect the new responsibilities of nations themselves, and of their constituents, especially powerful groups or large corporations with dominating influences.

For individuals, leaders, and corporations, both local and international, the constitution has to define limits to behaviors. Laws have to apply to large organizations, which are by no means necessarily private, or individuals or responsible. They must be liable for their behavior and effects. The nation has to reinstate the system of balances and checks, and government by law, with no exceptions.

Encourage Small Businesses for a Nation. Small business allows local employment, quick adjustments, and innovation. Small businesses built the United States in that country's first century; even today U.S. citizens hold high regard for small businesses, whose flexibility has provided lessons for big businesses. Nearly 98 percent of all Canadian businesses are small businesses. Small businesses contribute significantly to the economy of a nation, in innovation, in adaptability, and in job creation for women and minorities, as well as to distressed or depressed areas.

With most of the world's business being conducted by small entrepreneurs, it makes good economic sense for governments to implement policies that encourage small-business growth. The five ways in which government can have the most positive effect are by making capital more accessible, facilitating business education, promoting entrepreneurship, reducing regulatory burdens, and protecting intellectual property.

There are many things that go into creating a successful small-business economy, but the first might be the encouragement of entrepreneurs willing to start new businesses. For that to occur, citizens must be able to learn business skills. There are several ways in which governments can assist citizens. The community can create business incubators, as many universities in the U.S. do regularly.

Aside from lowering taxes to encourage business formation, it is important to reduce or eliminate those government regulations that slow business development or encourage business overgrowth. The simpler and faster the regulatory process, the greater the likelihood of small-business expansion. Any government that wants to encourage small business needs to produce laws that protect the innovations of entrepreneurs. Innovation is at the very heart of small-business growth, but if innovations are not legally protected, entrepreneurs may be unlikely to engage in the risks necessary to invent new solutions to social problems. Accordingly, policies that protect patents, copyrights, and trademarks help small businesses to flourish.

Protect Citizenship for a Nation. The rights of citizens used to be straightforward, perhaps because the definition of a citizen was more straightforward. As nations are becoming increasing complex and societies are becoming more socially and geographically mobile, the idea of citizenship needs reworking.

Nations have the right to identify their citizens, and to offer them benefits. The citizens of a nation have made many investments and sacrifices to promote their own livelihood or the health of the nation. But, there are problems. There is increasing tension between the rights and obligations of citizenship. The differentiation of equal citizenship into group-related rights and special legal statuses, for multiple citizens, refugees, resident aliens, and transient foreigners, puts a strain on the membership. There is growing ambiguity about the collective identity of a people in a nation with open government, and on the significance of citizenship as membership in a national political community.

Some rights are available to all people in a nation. The basic principle for this inclusion is stated by the 14th amendment of the U.S. constitution: "No State ... shall deny to any person within its jurisdiction the equal protection of the laws." Protection is equal for citizens and foreigners; and it derives from being inside the territorial jurisdiction of a nation. Stated in this way, this is a universal human right whose corresponding obligations happen to fall upon a particular nation, in this case the United States.

Although the nation has the power to limit citizenship, and the power to deny citizenship to settled first and second generations of immigrants, using this power will require constraint. Citizenship itself is not a right to anyone from outside. Although foreigners have access to basic human rights, they should not have the same access to citizenship and its benefits.

The rights of citizens might be better protected, and the power of public security authorities more restricted, under a law on identity cards. A law on ID cards for residents could maintain a balance between the function of public security and the protection of citizens' rights.

Create Long-term Ecological Planning for a Nation. For nations that intend to have a long-term existence, it is necessary to have long-term plans, and these plans should reflect the importance of an ecological perspective. It is crucial to know what habitats and resources are within a nation, to be able to delineate them, and to evaluate their health. A national survey is needed. The nation must begin the survey — perhaps the first that any nation would have — immediately. And, it would need to be a survey of every species, including viruses and bacteria, of habitats and cycles, including freshwater resources.

When the survey is complete, monitoring must begin to verify the kinds and extents of changes. When the ecological environment is evaluated and understood, the nation should link population to the carrying capacity of the land. It should link the consumption levels of the population to productivity of the land and to its ability to take and recycle the waste.

Survey All Resources: Measuring & Mapping Inventory. The first survey should start with the geological: Heat from the earth, thermal vents and volcanoes; geological formations, as well as specific elements. Then, an assessment should be made, of those elements which should be left in the ground, and how, of those to be taken, they should be taken, and how the land should be replaced to a desired condition.

Survey Air Water & Soil Resources. The air would be surveyed for circulation and quality, not just of the local system, but of the entire airshed, which requires information from neighboring nations. The integrity, productivity, and sustainability of natural ecosystems are intimately linked to air quality. Injury and death of terrestrial animals from airborne pollutants, such as metals and gases, have been observed since the 1870s.

Soil ecosystems are a sink for many air pollutants; wildlife that inhabit the soil environment are sensitive to soil contamination. Air emissions can cause reductions in soil organisms and shifts in trophic structures, such as insectivorous bird species. A reduction or change in decomposers can result in a decrease in litter decomposition and nutrient cycling. The distribution and abundance of salamanders may be influenced by soil acidity. In the United States, approximately 50 percent of the species of frogs and toads and 30 percent of the species of salamanders use ephemeral forest ponds for reproduction. These small pools and ponds can become acidic when they receive snow melt and spring rains that have little contact with the soil buffering system. The problem is that atmospheric monitoring data is almost exclusively urban. Local communities, schools, and individuals can join in the gathering of this important, wild data.

Water would be surveyed, not just amounts and the purity of aquifers and watersheds, but its regional cycles also.

Health is related to the fertility of the soil. Most every culture that has inhabited the planet has come to such an awareness — many after it was too late. We need scientifically-based data on soil resiliency, as well as on the relationships of soil characteristics and conditions to gains or losses in productivity and hydrologic activity. The maintenance of soil quality is one of the most important requirements for long term sustainability of the productive capacity of forest ecosystems. We must use tools and methods that provide early warning signals of impaired soil productivity.

Social and economic health requires us to regulate and monitor those management practices that have the potential to reduce soil productivity. Soil, along with climate, landscape morphology, species diversity and other factors, sets the limits on productivity within a biome through the flow of nutrients, moisture, and air supply to the soil builders. Soil condition reflects a wide variety of other variables, and is therefore a direct and key indicator of site carrying capacity and landscape productivity.

Productivity Resources. The productivity of the land would be measured. Ecosystems are dynamic: Trees and plants are the food sources for all other organisms. Almost every food web is ultimately dependent on the amount of plant tissue or biomass available for consumption. Most methods for measuring productivity are based on the repetition of biomass measurements at several points in time, with the increase in biomass representing the net primary productivity.

Productivity studies are very important for a number of reasons: They indicate a great deal about the dynamics of natural ecosystems; are of great value in agriculture and forestry, with domesticated and wild crops and resources; and, are useful in the study of plant-environment relationships through the application of bioassay techniques.

Productivity rates have kept the atmosphere functioning in a way comfortable to human beings. Carbon dioxide accumulates in the atmosphere, although deforestation and fossil fuel consumption increase the amounts. Reforestation could remove carbon from the atmosphere. We need about 7×10^6 square kilometers of new forest to store 4×10^6 tons of carbon annually, according to George Woodwell's estimates.

Estimates of the minimum vegetative area for the planet are more difficult to arrive at. Houghton et al. (1990) suggest that the minimum should be about what remained in 1990: About 5.3×10^9 hectares, or 40 percent of the land area, although the area remaining that year is not definitely known. Research is needed.

Monitor All Resource Use for a Nation. Monitoring should be a national effort, involving scientists as well as citizens groups and special organizations. I.F. Spellerberg defines monitoring as the "systematic measurement of variables and processes over time for a specific reason," such as ensuring that standards are met. Monitoring is the measurement of the parameters that define patterns that indicate health or change in an ecosystem or place.

There are different levels of monitoring, including environmental, biological, and ecological. Environmental monitoring is an umbrella for many activities, including climatic variables and geological processes; for example, the systematic recording of soil and air temperatures, humidity, air pressure are measured to predict long-term climatic change.

Biological monitoring is the regular, systematic use of organisms to determine environmental quality; that is, the state of the environment can be analyzed by how individuals react to pollutants. Biological monitoring has numerous subcategories, such as the biochemical, microbiological, epidemiological, and biotelemetry. Ecological monitoring is the observation of communities to understand long-term ecological processes, such as succession and maturity.

Before monitoring can begin, the objectives and data collection methods have to be nailed down. The purpose of the monitoring program has to be stated; the objectives have to be identified, for instance, the health of ecosystems. Health will probably be the first objective, followed by production or aesthetics. Health can be defined by a set of indicators of health, such as species, or patterns of health, such as stability or productivity. None of the indices that are measured are really adequate to define the health of an ecosystem, because they cannot account for the complexity, richness, and cycling that goes on in the ecosystem. For that reason, health indices are data that need to be resolved on the ground in person by someone who knows the history of the place and has a feel for it.

The basic medical definition of health used to be freedom from disease. Part of a new definition is resilience to stress — of course, there is good stress (eustress) as well as bad stress. Ecosystems respond to stresses in different ways, but usually through a decrease in the indices mentioned. The symptoms of ecosystem stress are major items to be monitored.

Scale of Monitoring. We have been monitoring stands and patches. Ecosystems are larger than patches; forests are larger than stands. We should be measuring at larger scales: Watershed, landscape, or biome. Although some of the technical tools, such as satellite imaging and GIS, are being used, many conceptual tools, such as the Gaia Hypothesis or global design, have been neglected.

Furthermore, we are measuring over short periods of time, a year or two, only to establish a growth rate or productivity. We should be measuring over centuries. We cannot use a short-term industrial approach to measure a few parameters and then pretend we know enough about an ecosystem to take a large percentage of it. Forests ecosystems, for instance, are created by slow processes that take hundreds or thousands of years. In their 36-year study of a 450-year old conifer forest in Washington, Franklin and DeBell (1988) projected that it would take the shade-intolerant Douglas fir 750 years to drop out of the forest. Maser points out how long it takes for coarse woody debris to decay (200-460 years). Soil formation takes millennia; rates can range from 50-100 years per centimeter. We need very long-term studies.

We need to be monitoring every part of the web of interdependence. Monitoring everything will give us a better grasp of the health and normal changes in a forest — especially compared to interference or degradation through improper use. Monitoring should be as comprehensive as possible.

Monitoring shows how an ecosystem changes and at what rates. It shows what areas of the system are critical for the functioning and ecological integrity of the system. It may give an indication of how to rank values in the system. It shows how use of the system can be balanced and how the use can be zoned according to degrees of protection and use.

There are a number of problems or limitations with monitoring. The processes of many ecosystems have not been researched, so there is no baseline to compare measurements with (it is difficult to measure change without a baseline). Human-caused disturbances have long-term, synergistic, cumulative effects that are hard to trace. Furthermore, with the complexity of the measurements and the limitations of laboratory facilities or labor; the costs for analysis for some chemicals and metals, may be prohibitive. The cost of investigating over a large area for a long time may be very high.

To some extent, these problems are the reason that monitoring is not a regular part of management plans. However, monitoring is crucial to understanding ecosystems. Until we understand how systems change and move around the landscapes, we will not know which changes are important and inevitable and which are the unhealthy result of human interference. Until we understand the changes, we will not be able to adjust our needs to limits.

Establishing a Context for Ecosystem Monitoring. The actual substance of the environment consists of patterns as well as things or individual species. The environment is generated by a patterning of the ecological ebb and flow of energy, substances, individuals, and species across a suitable landscape. The distinction between growing and declining patterns is not arbitrary, and can be arrived at objectively. And, the ecosystem environment is constituted of a large set of events that are objectively definable by specific outcomes.

The procedural consequences of these facts involve practical changes in the relationship: Between the earth and its species over space and time; between the earth and the collective ability of species to respond flexibly to situations beyond their normal patterning in the processes of adaptation and repair; in the flow of energy through the environment and its biotic community; and in the ecosystem's collective patterning of shared needs and governance which we will define as its health or soundness.

Taken as a whole, the process of observing patterns in these relationships leads directly to a fundamentally different way of perceiving ecosystem management. As Richard Hart mentions, patterns are the key to understanding the nature of a forest. In some ways, patterns are prior to things, in helixes, light, fields, and ecology. Paul Shepard and others have written that relationships are as real as the objects that result from them. Ecology attends the overall pattern of relationships, beyond the details.

The challenge to measuring and monitoring ecosystems is to address the patterns. But, the tools will have to be used in new ways in a new framework, perhaps with topology and holograms as metaphors (topology provides a mathematical model for processes, and a hologram provides a model for connected wholeness).

Based on a broader foundation, with more comprehensive values, measuring (mensuration) and monitoring need to address patterns of being in an ecosystem and not just a few commodities. One challenge is to identify the patterns and set up long-term programs to study them and relate them to sustainable use of ecosystems.

Protect Resources for a Nation
Once the extent of the resources is known, and once they have been categorized for use, they need to be protected. Protection may be temporary, for future resources or use, or permanent, for areas critical to regional and global cycles.

Minerals. Because many minerals are so unevenly distributed around the planet, many nations have few while others are well-stocked. Some minerals, such as copper, are circulating at a higher volume than what remains to be easily removed, such as massive amounts in low concentrations in seawater. The large copper and iron mines, that require the entire landscape to be removed to reach the seams, should be closed so that the landscapes can be restored. Recycling programs are more efficient for those minerals at this point.

Fossil fuels have also reached the point of diminishing returns. At our high level of use, new discoveries are only extending our useage by months. Usage could be reduced quickly through conservation. New methods, such as horizontal fracking, need to be evaluated and regulated, or abandoned — fracking requires too much freshwater and its wastes are too expensive for a community to clean up.

Allocate & Ration Water. Fresh water on the planet is unevenly distributed; 20% of fresh water is contained in the US great Lakes, and another 20% is in Lake Baikal in Russia. The rest is in rivers, aquifers, and lakes, for the most part. Humans have made large-scale alterations in water and material fluxes in rivers, causing erosion, and adding quantities of minerals and nutrients. We have dammed most rivers, some rivers many times, to get water for agriculture in general and irrigation for dry areas. We are draining aquifers for the same reasons. Unfortunately, the health of aquifers and rivers is declining across the planet. Many rivers no longer reach their deltas to the seas or ocean.

At this point we need to start addressing the overuse, as well as the interference in every ecological aquatic system to avoid catastrophic collapses. We need to allow aquifers to recharge, which may take from months to thousands of years. We need to restore river systems. And, we need to recognize the value of water and charge more for it. But, even as we increase the cost of water, partly to pay for restoration, we need to allocate much of the water to the system itself. The health of the system has to trump all other uses; no river equals no water.

Here is how allocation could play out. In a river, such as the Tigris in Iraq, the annual mean discharge at Mosul in 1996 was about 600 cubic meters per second. Assuming that 410 were needed for river maintenance, that leaves 190 for human use. Some uses are critical, such as drinking water and some irrigation, and other uses are luxuries. Drinking is critical for people and livestock; estimate that at 40. The demand for irrigation may be as high as 200 — in fact the demand is such that the regional aquifers are being drained for irrigation now. Given the limits of availability, irrigation will have to be limited soon. So, there is no water left over for luxury needs at all. If there were water left over, it could be allocated by category, and then rationed out by percentages for each use. Marginal cost pricing of water would allow the price to float according to demand. Water is too valuable to be sold cheaply, wasted or used unwisely.

Save Important Ecological Areas for a Nation
Saving used to mean isolating the areas entirely from human use and interference. But, human aesthetic use and human basic use, as done by archaic cultures, is often part of the process itself. Saving has to mean regulating use, but this is more concerned with reducing consumption below certain levels and not allowing large-scale use for external requirements.

Our species has been shaped by the earth. The desire to save forests, wetlands — all natural ecosystems — is an expression of deep human values (or perhaps a more basic survival instinct). Experience of wildness lets us capture some of our own wildness and authenticity. Our emotional response to the unfathomability of the ocean or luminosity of the desert is an expression of aspects of our fundamental being that are still in resonance with these forces.

We have learned that we cannot just save big trees or waterfalls or geological formations. We have to save the system and process that produces big trees or spectacular forms that we like. Size is important. Thomas Lovejoy and others have made parallel arguments on species loss, leading to the idea of a minimum critical size for ecosystems. The minimum critical size is greater than the areas suggested by the species-area curve because of three reasons, Lovejoy says: Species are identified by individuals, not breeding populations; the species-area relationships are derived from extensive habitats rather than the fragmented ones; and, higher trophic levels, such as tertiary predators, may be excluded from smaller areas — that is, all species are not equal. As an ecosystem is first reduced in size, it can still maintain its characteristic composition and species diversity as a self-sufficient, functioning whole. However, as it is fragmented and impoverished, more sensitive species drop out and its function is impaired. At some point, as the size is reduced below the minimum, the integrity of the system is compromised and the system collapses.

We have learned that shape is important, also. We have to protect interior species and conditions as well as exterior ones. Edge effects started out as being centers for diversity. Now, many edge effects are considered to be destructive. In the 1940s, it was thought that edges should be increased to provide bountiful game crops. Edges have proven to be good for both game and wed species. Edge effects, however, are detrimental to populations adapted to ecosystem interiors. Too many edges can reduce diversity at local and regional scales. We have to anticipate changes at global scales.

Restore Natural & Wild Areas as Necessary for a Nation
Restoration is one of the major ways to ensure the survival of species, habitats, territories, and ecological systems. Restoration projects have the potential to save entire ecosystems. Agricultural systems might save domesticated species. Restoration ecology, as a new discipline, tries to address the difficulties of deciding the goals and means of restoration, considering the lack of information about original systems and the loses of component species during the degradation or destruction of the wild ecosystems. The discipline attempts to create self-maintaining neopoetic systems characterized by complexity and diversity.

A holistic science such as landscape ecology addresses the overall patterns of large-scale ecosystems, considering the biogeochemical, atmospheric, and hydrological cycles in relation to the shape and extent of individual landscapes. Landscape ecology can identify candidate ecosystems for restoration, as well as for preservation, conservation, or reservation; it can identify patterns to preserve larger functional islands.

Restore Forests. A crisis science like conservation biology or ecoforestry can make recommendations which would preserve diversity and complexity in forests, and would avoid numerous extinctions. The first recommendation is to stop logging old growth and mature natural forests, Then, promote cutting practices that respect the productivity and complexity, leaving snags, logs, and many-aged forests. Grant timber leases that are contingent on the maintenance of the productivity and diversity of the land. Reduce fragmentation through the design of forested areas, taking into account the genetic diversity of the trees, catastrophic conditions, minimum viable populations, corridors, and edge effects. Other recommendations are to: Stop constructing new roads; close and revegetate old roads; restore clearcut areas; replant with native species; restore damaged streams and wetlands; restore natural connections, such as corridors and canals; and recommend that reserves be made large enough for minimum viable populations and minimum viable ecosystem areas. Restoration areas, which are set in a pattern by human activity, but may not need further intervention. All areas need plans.

Rewild Grasslands. In Africa, elephant, lion and other carnivore and herbivore numbers need to be returned to preharvest levels (or before 1500 CE); perhaps 10 million elephants. Elephants are ecosystem engineers, reshaping their landscapes. This results in changes of species. Lizards lose out, but antelope and zebra prefer elephant savannas; and lions and other predators prefer antelope and zebra. Elephant droppings are rich in undigested plant seeds and elephant-made waterholes help other animals. Elephants tamp hard paths that other animals used. Elephants can make caves to mine salt; the caves are later used by leopards and bats. Elephant lands need to be identified and set aside, then linked with corridors between them.

In India, elephant and tiger populations need to be returned to earlier levels (and protected from capture and use), probably to 2 million elephants and 100,000 tigers. Since over 90% of their habitats have ben converted to farming, much of the habitat needs to be restored and reserved, with some sharing with farmers.

In Europe, native mammals included aurochs, bison, boar, chamois, red deer (elk), horse, ibex, moose, and reindeer (caribou). Poland has reintroduced European Bison into the Bialowiea forest. Spain has some wild horses. Italy and France have ibex. Scandinavia has Sami herds of reindeer. The aurochs and other unknown animals are extinct (and possible candidates for genetic reconstruction). Abandoned areas are 'available' for the reintroduction of wild animals, such as wolves in Bulgaria, who could shape the systems into a more natural configuration. Even small single hectare lots could be left for smaller mammals who are gradually making their way into cities on their own. Larger expanses could be made into more formal parks. Large mammals would be introduced from remaining stocks. Mammals from other continents, as functional equivalents, such as African rhinos to replace the extinct variety in Eastern Europe, would be added to the mix. In some areas, rewilding would begin with trees, then beavers, lynx, and wolves. In others, storks, pigeons, falcons and bats.

For North America, Dave Foreman, Reed Noss, and Michael Soule have suggested rewilding the Great Plains, starting with cattle and bison. Cattle could be run in large numbers as equivalents, until wild species could be introduced. Mammals from other continents, as functional equivalents, such as camelid, elephantine, and rhino, would be introduced species. Eventually the system would be more balanced and self-renewing (and self-managing, of course). These areas could be left by benign neglect or set into motion with sophisticated management; they could be connected by corridors, probably abandoned railway lines or decommissioned roads and highways, which would link the rewilded patches and allow migrations (and genetic flows) in the whole matrix, which would include large areas for farming. A balance between edge and interior in an overall wild matrix is important for large species and for interior species. The apex predators, such as wolves and bears, help balance a system by limiting the cascade under them, all the way down to the native tall grasses, flowers and herbs. The Great Plains might become a more dangerous area, but that might be welcome by many.

Conserve Use of Resources for a Nation
Natural resources were originally defined as objects provided by nature for human use. This concept has been expanded to include minerals, wildlife and people. The idea that everything should be managed is based on an extreme belief that nature is a resource to be processed. Furthermore, management is self-perpetuating and self-justifying. The objective of resource management is not to strengthen its defenses or funding, or to increase quality of life for affluent people in overdeveloped countries — it is to adjust the use of resources for the needs of current and future generations.

In short, conservation management is based on economic objectives. And, as Leopold pointed out, the weakness of relying on economic motives is that most members of the earth's community, such as wildflowers and songbirds, have no economic value. Yet, all the members of the community contribute to the integrity of the whole, which is vital to maintaining what we do consider important. Those beings with no economic value are ignored, or worse, labeled as weeds or vermin and destroyed so that crops and animals with short-term advantages for human ends can be substituted.

Most conservation strategies are completely anthropocentric, from saving hunting grounds in the middle ages to allocating resources this year. This plan proposes ecosystem conservation, which protects entire biotic communities: Genes, populations, species, habitats, associated traditional human cultures, and all the processes and interactions.

Conservation parks are a way to keep resources. Conservation parks are areas set aside for multiple use of resources without interfering with the operation of the ecosystems. Research may be conducted to answer questions as to whether the park is big enough and shaped correctly to constitute a proper habitat for its inhabitants.

Limit Use of Common Resources for a Nation. Common areas need to be preserved. Common areas can be used so that they are preserved in character and function. This can be done through the following limits: Limit access; limit use by large vehicles or houses or buildings; limit harvest; and, limit grazing. This may mean limiting the size of flocks or herds to allow other wild grazing animals. Hunting, grazing, and agriculture provoke large ecological disturbances.

In general, mammalian grazing promotes regrowth and the movement of seeds. Bison and prairie dogs were responsible for much of the character of the American plains, but cattle have led to fencing and the eradication of pests, which leaves the ground overeaten and muddy. The values of keeping healthy ecosystems, such as temperate grasslands, are far more than the perceived losses from limiting domestic animals.

Limit Use of Slowly Renewable Resources for a Nation. Slowly renewable resources can still be used, but with respect to their extents and rates of renewal. In general, we need to: Limit takes to the renewal rate; limit the styles of taking; and, require replanting. Charge the cost of use by replacement value, which may be substantial.

Limit Aquifer Use to Recharge Rates for a Nation. First, we need to determine or survey aquifers, then determine recharge rates, then determine the value of water and the cost according to value. To supplement the use of aquifers, plan for the collection of rain or wastewater to reduce demand. Monitor all water use. Treat the aquifer as the principle and water flow as interest.

Use & Limit Wild Harvest for a Nation. If large areas are to be allowed to operate with natural processes, then large areas have to be limited regarding human use or conversion. Perhaps, the human use of some wild lands, within limits, can continue, but it must be part of a total change, unlike permaculture and other strategies which can be practiced as part of an industrial lifestyle. Paul Shepard argues that 75% of the land area should be left wild in a techno-cynegetic society, so that human beings can return to the hunting and foraging lifestyles that shaped our species.

Even if a return to this style of life is not possible for most people in most nations, some ideas can be taken from it. Wild animals could be harvested for wild food, if it can be done with minimum disruption to their breeding and movement.

Create Large Conservation and Wilderness Areas for a Nation. Create large conservation areas, such as the Wildlands Project recommends for the Rocky Mountains and Great Plains in the U.S. to the Appalachians or Sonoran desert. The goal of the Wildlands Project (Dave Foreman et al.) is to set aside approximately half (50%) of the North American continent as 'wild land' for the preservation of biological diversity, by creating 'reserve networks' across the continent, which would be composed of cores, buffers and corridors. The primary characteristics of core areas are that they are large (from 100,000 to 25 million acres), and allow for little, if any, human use. The primary characteristics of buffers are that they allow for limited human use so long as they are managed with native biodiversity as a preeminent concern.

Many nations need to have forests and wildlands restored. Mesopotamia could benefit from such a wilderness (and restoration) plan. Many European nations are setting aside 6-9% of their areas for rewilding. China has been setting aside many large wilderness areas.

Ireland is a nation that was heavily forested, but now is one of the least wooded lands in Europe. Richard St. Barbe Baker calculates that the minimum forest cover for safety is a third of a country's land area, far more than the current cover in Ireland. Furthermore, he states that about 22% of a farm put in shelterbelts could double the yield of the farm.

One recent proposal recommends replanting and restoring over 30% of the land area in forests, by concentrating on lands with poor soils, on hills and along corridors that would connect the forests across the island in one linked network, which would allow plant and animal movements to resume around the edges of agricultural fields (See Figure 4).

Management would include a network of Monitoring Sites where different management approaches in various woodland types could be assessed, as well as professional conferences, training courses, workshops, and manuals and guidelines. Management has to address threats to forests. Invasive exotic species, such as rhododendron and laurel, are major threats to native forests. They must be controlled or eliminated so that they do not prevent natural regeneration.

Conserve Places and Species for a Nation. We have always tried to exceed the physical and biological limits of places rather than recognize them and be guided by them. Most places exist in a uniquely identifiable ecosystem, with recognizable boundaries and a unique history and character. Modifying such places can change or diminish their uniqueness. We can conserve places and species by respecting their limits and limiting our exploitation.

Privatize Communitize or Nationalize Resources. Some resources might be saved through a program of privatization. Perhaps we should privatize where applicable or reasonable. When something is privatized, however, the profit or loss accrues wholly to an individual or group, such as a corporation. Privatization is wonderful for individuals, farmers, and small businesses. It inspires people to work hard, save, and keep their efforts healthy, especially if a forest or farm is the basis for their efforts. Privatizations sandwiched Chinese efforts to grow crops efficiently and were found to be preferred by farmers. Privatization is often limited by scale; it works at small scales where the individual knows the details of an operation and profits from their own efforts and restraints.

However, the privatization of large holdings to individuals and corporations does not seem to have worked well. With no sense of stewardship, with a focus on short-term profit, and with little understanding of ecology, land gets exhausted quickly, then used or sold for less profitable ventures, such as running cattle or building houses, which further degrade it. When land is put away from use, or rather for the exclusive use of one person or group, who does not always use it, then it may be conserved. Thus, kings or groups of monks quite often saved trees and forests by putting them behind walls and limiting access. It worked by accident. Larger patterns of land and things should be communitized, that is owned by whole communities, who would enforce rules for use and restrict access. Communities, like monasteries, are managed usually for the long-term.

For whole regional systems, with watersheds, airsheds, and landscapes of ecosystems, nationalization is a better strategy. The nation can set aside whole systems. The nation can better balance resources with wilderness areas. The nation can take a larger view than the community, although both of their perspectives often are longer than that of an individual or business group.

Implement National or Regional Ecological Goals. Once goals are identified and agreed upon, they can be implemented. National governments have been comfortable with short-term economic goals and a few ethical goals, but have neglected goals dealing with the ecosystems and climates.

Anticipate Climate Change. The climate is changing. The changes have been documented. It is just that politicians, and most of their constituents, are slow to react to these average changes, which after all seem slight, compared to the extremes of weather. To a few the changes are threatening, not just to low-lying islands and a few corals, butterflies and trees, but to civilization. Global warming could lead to reversal of ocean currents and other unpleasant, deadly effects. What is the solution for climate change? Use alternative energy sources? Possibly, energy use could benefit from a high-tech solution. A massive development program, like that for the atom bomb, and with the same urgency, could develop and implement alternative energy technologies.

There are many less technical actions that could combat climate change. Planning could always consider greenhouse warming. CFCs and greenhouse gases could be phased out. Contributing agricultural problems could be corrected. Nations could consider full social cost pricing of energy, where the polluter pays. Deforestation could be halted; reforestation projects could be started. Biodiversity losses could be slowed or reversed. Efficient use of water could be increased. Aquifers could be protected. The easiest solution might be to reduce consumption and populations. Nations could share data and participate in international projects.

Identify Ecological Goals for a Nation. Regional or national goals are appropriate for bioregions and isotopes (literally meaning similar places). The number of goals decreases as the scale gets larger.

- Zone ecosystems at the landscape level for preservation, conservation, or selection use
- Maintain ecological and evolutionary processes in healthy landscapes
- Diversify the institutions that deal with ecosystems; relate people diversity to ecosystem diversity
- Reduce fragmentation of landscape patterns with responsible use
- Accept and plan for natural change, disturbance, uncertainty, ambiguity — it's part of the process
- Restore forest cover in North America to pre-European levels
- For the U.S., replant 142 million hectares of forest lands — up to 438 million hectares (the approximate level of cover in 1600)
- For Canada, replant over 80 million hectares of forest, up to 530 million hectares
- For the Northwest (Pacific Northwest coast forest, replant up to 47 million hectares
- For the Northeast (northern hardwoods), replant up to 11 million ha
- For the Southeast (Oak-pine forest), replant up to 129 million ha
- For other ecoregions, replant to a high percentage of 2000 BC levels, especially around the Mediterranean
- Relate population and consumption to ecosystem productivity — without such limits, forests will eventually be destroyed

Set Cultural Goals for a Nation. Cultural goals are more complex and challenging. Stabilization would be a good first step.

- Stabilize the cultural environment
- Stabilize population
- Encourage cultural education and history
- Encourage teaching of the local language
- Allow renewal of critical customs
- Work to establish equity between groups
- Allow more opportunities for fulfilling needs and generating wealth.

Long-term Economic & Political Planning for a Nation. Human interactions can become more violent as a result of competition, inequity, and limits, requiring more planning and social control. Local planning often ignores limits and carrying capacity, long-term deficits and problems, other species, and ecosystems. Local planning is limited to a two to four-year cycle. Regional and global planning are virtually nonexistent. Crises sciences, such as conservation biology and ecoforestry, plan at the landscape level, and economic and political planning need to plan at this level also.

Planning in general means deciding on goals to be achieved in specific situations. For central planning by a federal government, the goals are usually small and not comprehensive, such as a forest cutting level or a single species preservation, which usually end up being a compromise in cost-benefit analysis. Central and most other types of planning tends to neglect or dismiss the distribution of negative, uncertain, or nonmonetary effects, which are characteristic of much of nature. Furthermore, we seem to have no mechanism for developing long-range plans. Certainly, there seems to be no way to deal with long-term, slow catastrophes, such as deforestation or climate change.

Most plans address *problems*, such as building roads. Everything else, from employment to pests, is also considered as a problem, and not a direct effect of the cultural implementation of some pattern. Most plans are also development plans that are comprehensive in the sense of seeking to meet all needs of the public, agriculture, and industry. Development plans tend to call for the eventual development of *all* resources in an area. Development is more concerned with an assembly-line model—simple, isolated, efficient, and easy to maintain. We become remote from, and indifferent to, the system that supports us. We acquire unrealistic images of the world and harmful values and then make bad decisions based upon them.

A one-world planned economy is an even greater threat, being based on unlimited industrial production, unlimited commodity consumption, increased exploitation of nature, and the free flow of resources and labor across cultural borders. This kind of planning requires the abandonment of local controls on development, trade, or lifestyles. Planning is thus characterized by a utilitarian globalism that denies value to the systems that support it. As a result of central planning, the patterns of life have become the products of market forces operating in a sterile abstract order.

We have not developed qualitative indicators of ecological health or quantitative measures of social health, much less an ecocentric view that would value preserves of nature for themselves. One solution to many of these problems is a reduction in scale for everything from ecosystem use to management units, with local controls and local use primary. Management costs increase with the size of management units; more levels of human hierarchy are required to deal with problems; decisions are slower, and the people who make them are more remote from the site.

Obviously, a plan should consider the whole system. Human needs should be designed for an optimal fit within the limits of the system. Ecological planning considers the health of the system, which is based

on intimate knowledge of the system. Direct observation and traditional knowledge yield far more 'information' about the societies of plants and animals than autopsies or mathematical models. A whole-system, comprehensive plan would proceed in stages:
 1. Identify the place within its natural boundaries. Most places exist in a uniquely identifiable ecosystem, with recognizable boundaries and a unique history and character.
 2. Calculate the optimum amount of wilderness to preserve the natural cycles indefinitely. If the current wild area is less than the calculations, restore the difference and set it aside as a preserve.
 3. In the remaining area, zone areas for appropriate use, including conservation and artificial areas (roads or cabins).
 4. Identify the resources needed for human use, including raw materials and the productivity of the areas. This productivity can be used to calculate rational exploitation.
 5. Apply cultural modes — in style, values, and technology — to set limits on technology and population in the area to be supported.

We would examine the natural and cultural histories of a place, as part of our comprehensive ecological plan, which is actually a deductive, synthetic, conceptual model based on data generated from research on biological productivity, the rates of resource use, cultural valuation, minimum wilderness preservation, air and water quality, genetic minima, nonrenewable resources, appropriate technological innovation, the importance of cultural frameworks, adventure, research, beauty, uniqueness, and other intangible experiences.

Planning is not meant to be a finished work of art — it has to be a changing, adaptive pattern that reflects our participation, understanding and use of the system. Each activity needs to be fed back into the process of updating the plan. Implementing the plan would then result in adaptive improvements.

National Planning to Survey Human & Cultural Needs. Abraham Maslow listed a full hierarchy of human needs: Physical needs, such as food, shelter, and clothing; 'safety' needs, including law, order, and security; 'psychological' needs characterized by belongingness and love; 'esteem' needs, such as strength, self-sufficiency, competence, freedom, attention, and prestige; and 'self-actualization' needs, like self-actualization, achievement, and creativity, which develop after other needs are satisfied. Of course, human needs are based on the health of the earth. Human needs extend to include a foundation of wilderness. Human systems depend on natural ones, for recycling of wastes, water, and air. But, as human growth tends to be logarithmic, so does human need, and need shapes facts, like the kind and quality of resources. Granting this, the need for wilderness is as much a fact as the need for food.

Surprising things can be discerned about living a good life: Once basic needs are met, for food shelter, respect, and confidence, then happiness is not increased much by more material goods or money. The things that make people unhappy are when the higher needs are not met. Lack of love or security, lack of communication or appreciation, lead to unhappiness. People try to balance their needs by acquiring more money or things. The effort to balance may result in distortion or violence.

The society and the culture have to be meaningful, as well as secure and equitable, with low extremes between wealth or status. Similarly, there is a hierarchy of needs of a culture: To be grounded in place, to be secure or partly isolated; to have a dynamic order, with human health; to be complex and sophisticated, with checks and balances; to be comprehensive, to allow change and diversity; and, to have and manage adequate resources. Alas, sometimes cultural needs can be perverted, and the needs are almost completely defined in terms of commodities.

It is human values that ultimately determine what, where, and how we monitor management activities. Therefore, it is imperative that cultural values and natural processes be determined and monitored simultaneously. In this way, we can begin to understand which natural processes we use, or favor or discourage, in order to benefit our current value system. Monitoring and evaluation of indicators provides necessary information for understanding the human dimension as it relates to ecosystem management. What makes something valuable is not in its own properties but in its relation to the personal preferences of its perceivers. However, the self-value or usefulness of a thing, being or species does not rest entirely in its appearance but rather in its existence.

National Planning to Monitor Human & Social Resource Use. The Earth Summit's 'Agenda For Change,' states that a stable human dependence on natural resources is key to the protection of forest and range lands. Social and economic data is often compiled by standard political boundaries, not on provinces, sub-provinces, landscapes or project sites. Thus, socioeconomic data is not strictly comparable or usable at the sub-province, landscape or project site levels. However, the indicators can be either described or mapped onto geographic information systems (GIS), providing an opportunity for integration with other physical, chemical and biological indicators.

For instance, the indicator 'demographics' provides information on where people settle, how many there are, and what they do. This has a profound effect on the environment and its sustainable use and management. Information gathered from non-national sources would assist us in making sound decisions that address the diversity of the local and national human populations. Data may include age distribution, in-migration, percent of population using resources directly (including recreation visitor days and employment), and where people settle—their distribution, and number including rural interface zoning, roads, type of industry, public services, communication lines, and community activities.

For cultural influences, information gathered through archaeological sources and studies provides information on the historic uses of the surrounding landscapes, and of associated human activities. This data may include ethnographic information and historic use patterns.

National Planning to Promote Efficiency of Resource Use. Data on local and regional economic health would help us to predict long range trends and landscape use patterns. Data may include real unit costs and price changes, income tied to resource activities, profitability of management, payments to government, percentages of products recycled or reused, changes in the distribution of employment and income, employment patterns tied to resource activities, rural interface land values, types of commerce, welfare payments into the community, and the economic base. Data may include various land-use patterns such as percent of land base set aside in protected areas, dispersed and concentrated recreation areas, grazing allotments, mining activities, vegetation or animal usage (game management areas, firewood, timber, special forest products, fishing spots, water withdrawals), and community contacts (retirees, business people, school students faculties and staff, community clubs, and tribes).

National Planning to Anticipate Technology Integration. Technology can be broadly defined as the material entities created by the application of mental and physical effort to nature in order to achieve some value. In its most common use, technology refers to tools and machines that may be used to help solve problems. Technology is a technique that lets us use resources to produce products and solve problems; this technique feeds back into culture. Due to the increasingly widespread use of ever more complex technologies and their frequently unintended consequences, problems may arise in their use that are unknown or only partially addressed. As tools increase in complexity, from knives and levers to computers and space stations, so does the knowledge needed to support them. Complex modern tools require libraries of information that has to be continually increased and improved, then spread and understood.

The use of technology has a great many effects; these may be separated into intended effects and unintended effects. Unintended effects are usually also unanticipated, and often unknown before the arrival of a new technology. Nevertheless, they are often as important as the intended effect. The combination of techno-cheap goods and complex tasks can lead to sweatshop slavery and unsolved wastes. Some problems are solved, but new ones are created by the unconsidered use of the technology — problems such as toxic waste. The most subtle 'side effects' of technology are often sociological. They are subtle because the effects may go unnoticed unless carefully observed and studied over large areas and long periods of time. These may involve gradually occurring changes in the behavior of individuals, groups, institutions, and even entire societies. A nation needs to address the effects, all of the effects including unwanted ones, of technology on business, culture, management, and the environment.

The implementation of technology influences the values of a society by changing expectations and realities. The implementation of technology is also influenced by values. There are major, interrelated values that inform, and are informed by, technological innovations. These realities and expectations may alter the world view of a culture, especially as regards efficiency, bureaucracy, or progress.

A nation has to define the goals of technology, from simple technique to the simplification of chores, as well as to consider the concept of nonharm. Then, it has to analyze the results of technology, not just the mass production of things, but the effects on human health and behavior. A nation has to discern what is missing from an application of technology, whether it is human scale or appropriateness. Then, a nation has to decide how to manage the technology.

Many new technologies are not managed for their perceived benefits or losses. Nanotechnology, for example, has the potential to clean chemicals and viruses from the human body, but some minute nanos could pass through various barriers in the body and interfere with brain processes or environmental cycles. Technology can be used to develop renewable energy and restore damaged environments. Technology can be used to promote localized energy industries, with solar, wind, hydro, tidal, or biofuels. But, technology has to fit the scale and tempo of the environment. Technology has to be sustainable and replaceable.

Technology needs to be integrated into society. It needs to be made appropriate to the goals and desires of a culture. The notion of appropriate technology, however, was developed in the twentieth century to describe situations where it was not desirable to use every new technology or those that required access to some centralized infrastructure or parts or skills imported from elsewhere. The ecovillage movement emerged in part due to this concern.

Technology needs to be integrated into the entire environment. Some technologies have negative environmental effects, such as pollution and lack of sustainability. Some technologies are designed specifically with the environment in mind, but most are designed first for economic or ergonomic effects. The effects of technology on the environment are both obvious and subtle. The more obvious effects include the depletion of nonrenewable natural resources, such as petroleum, coal, ores, and the added pollution of air, water, and land. The more subtle effects include debates over long-term changes, such as global warming, deforestation, natural habitat destruction, and costal wetland loss.

National Planning to Encourage Vernacular Design. Ivan Illich points out that the term 'vernacular' is an old technical word used for Roman law, having to do with 'free' access from a commons or free ownership, such as getting offspring from a donkey already owned. As Illich further points out, the vernacular is the opposite of a commodity, that is, something for which one has to pay. Illich suggests, in the tradition of Schumacher, to use the term 'vernacular value' for those things made by individuals and groups that are not destined for the market. These values imply two kinds of technologies, however, one for mass-producing commodities for the market, and the other for the production of personal or community goods that remain in the community. Illich questions the use of the word production, but it is used here in the large sense, as in the production of edible matter by plants.

In a commodity-centered society, ethics, politics and justice are reduced to equitable distribution of commodities, according to Illich. But, the access to resources for vernacular and commodity productions, with lessened economic domination of the latter, can be equitable, without reducing politics to simple access. Illich suggests that there are two choices for educating in a technological society, regardless of whether the energy is hard or soft: (1) People need further education to build solar collectors, or (2) education needs to be made transparent by the engineering of tools. Illich's idea of the vernacular describes a great divide, not just of production and education, but of language, with an every-day vernacular language and a technical, formal language.

Vernacular architecture is a way that cultures express a shared heritage in patterns of construction of shelter. It is a term used by the academic architectural culture to categorize structures built by nonprofessional or untrained builders. Although modernity should not be cause for exclusion, true vernacular is most apparent in the archaic world where indigenous populations produce their own shelter based on traditions of using locally available materials. This architecture can include a wide variety of structures, though domestic and agricultural buildings are the most common.

Another distinguishing feature of vernacular architecture is that design and construction are often done simultaneously, onsite, with nonmanmade materials. And, many of those who eventually use the building are involved in its construction, or at least have direct input in its form. Vernacular building shapes, construction techniques, and other characteristics are often generated from centuries-old local patterns. These patterns continually change, incorporate new technologies and values that perpetuate cultural norms. Vernacular buildings can reflect sophisticated adaptation to the environment and needs.

We have forgotten that the whole purpose of architecture is not efficiency or luxury, but is time and space for living well. For living well, efficiency is unimportant. Most importantly, vernacular design and building allows people to participate in meeting their own needs and expressing their cultural values.

Redesign and Replace Cities
Many cities were placed in flood zones or on fertile agricultural soils or in wild areas. Cities need to be matched to a proper environment; for some old cities, this means rebuilding them or moving them. New cities can be built for their environment. Some of these new cities could be arcologies, which are ecological cities with a very small footprint, internal transportation, and close to agricultural and wild areas, without impinging heavily on them.

With arcologies, and their urban ecologies, the city could change its relationship with nature; an arcology is a good solution for an urban culture, as it solves the problems of waste, resource-use, scale, obsolescence, and segregation. The placement of a new city is important. It should be sited away from flood plains and other geological threat areas. It should located be away from fertile fields, as well as from wetlands or wilderness areas. Surprisingly, the use of a single ecosystem for living, working, growing food, moving around, and creating things, disposes of many of the problems from the artificial separation of human and nonhuman systems. The city becomes an integrated ecosystem that may be more self-reliant and stimulating.

The success of life in an arcology, as Paolo Soleri contends, depends on miniaturization, where a prodigious number of overlapping mechanisms are packed into a small space. These mechanisms persist through built-in regulation circuits, but the systems are open enough for novelty. As an emergent form of life, traditional and modern cities have packed patterns and increased their intensity. Arcological cities have the potential to increase it still more, while decreasing their impacts on the environment.

Arcological cities might be able to reduce or strengthen connections in a more flexible pattern. Of course perception is a large part of patterns. And, we perceive the direction as being towards more complexity and more integration until we have megacities in a global society, coordinated on several levels, within a more complex biosphere. An arcological pattern would allow local cities to maintain their diversity be optimally integrated into new global forms. Self-sufficient cities would allow residents to start thinking locally and acting locally, within the ecological designs of the cities and within global constraints.

With an arcology, the city can change its relationship with nature; an arcology is a good solution for an urban culture, as it solves the problems of waste, resource-use, scale, obsolescence, and segregation. An arcology is a monumental design, which appeals to human desires for creating monuments. And, it could be a work of art in itself, which would satisfy the human aesthetic sense.

The size of an arcology could allow great diversity in retail stores and job opportunities, especially those related to running the arcology. The shape will allow optimum use of sunlight. Common areas will be designed in. Since arcologies would be highly connected, there is no reason that all of them have to be of a heroic scale. They could be mini-arcologies linked closely by rail, not totally separate but clumped in the area, like a mountain range.

Arcologies could be a better response to drought than traditional cities. Cities have been promoted as being efficient concentrations without size limitations, by Stewart Brand and others. But, if cities were organic, like an arcology, then they might have organic limits to overall size and density, as do bee or termite colonies.

With new low-carbon techniques and alternative energy collection and generation, the construction of an arcology would not have to rely on traditional technologies, such as concrete manufacture or energy generation from fossil fuels, which contribute to environmental problems, such as carbon dioxide release. The start-up costs would be relatively high, possibly

$20 billion or more. Energy and agricultural savings, however, should return that amount within ten years, at an optimistic guess.

An arcology creates a separate ecosystem like an episystem that participates in the cycles of nature, but spins new fabrics and elements, as well as new ideas and adventures. The whole premise of arcological design is that it has to be balanced with wild ecosystems, that is, nature has to be woven in with the city or vice versa. The scales have to match. They have to be balanced. That does not mean that every part of the city has to have wild tendrils in it. We can still have solid concrete and glass. Perhaps some of the parks could be truly wild. Working that out will take experimentation.

What should cities look like? Could we place them on or around mountains? This would solve the problem of interior space and lighting; everyone could have a view. A construct as large as a mountain could support the vegetation of a mountain, from grasses and trees to lichens and annuals. More importantly, the vegetation would contribute to atmospheric and edaphic cycles. It would attract complements of birds, insects and animals. With forests and streams, as well as agriculture and fructiculture. Arcologies complete the idea of the city and make it ecological, sustainable, frugal and exciting to live in.

Paolo Soleri has been designing arcologies for over 60 years, many of them set in Arizona, but others proposed for Europe and earth orbits. He has been building the first arcology at Arcosanti. His most recent design is the Lean Linear City: Arterial Ecology (presented in the book of the same name). Other arcologies have been designed for Africa, Antarctica, China, Canada, Greenland, and the US (Wittbecker et al. 2006 in *Redesigning the Planet*); for instance, the Oregon Tree Arcology, the underground Palouse Arcology, the Floating City of Venice, the Antarctica World Pyramid Arcology (for the UN or GU), and the Mumbai Shell Arcology.

Employ Wild Design
Traditional design has done well with cars and houses. Ecological design has created sod-roof buildings and parks. But they have not attempted large systems, such as urban areas or global spheres. Wild design, as a form of global ecological design, can address these more complex levels. It is as subversive as any art or design; it can overturn some wicked design problems. Global ecological design is a wild way of thinking that can mesh its approach with the wild planet, using an ecological perspective, systems understanding, participation, and standards of knowledge. It is wild design.

Wild design has to coevolve with wild nature in the planet. Wild design has to be adaptive. Within a framework for planning, we need to be able to accommodate uncertainty, the unplanned and the unimaginable. The framework will be incomplete; it will lack detail and definition, but it can be used by all the participants to coordinate actions within physical, ecological and mutual constraints.

Wild design works out challenges and problems in an artistic way. Art is wild. We cannot control the effects of art, or even anticipate all of them. We cannot anticipate the changes it might make. That artistic way is a wild way of thinking and can mesh with large-scale design better than a simple technical approach. Design needs to become wild. Wild design is not human-centered, as most all design in the past has been. Wild design is based on radical ecology—it is the push beyond human interests to consider the character and patterns of ecosystems. Not to subvert or interfere. We can guess what the system 'wants' or how it is developing with reference to its past behavior. We actually know what it wants: To exist and to regenerate. We need to create the conditions for the system to flourish. And, if we use any of it, that use has to be limited to that level of productivity that does not interfere with the survival of the system. We are reintervening in a natural system at different levels, rather than using or interfering for human benefit exclusively.

Wild design is modified in this sense to use power with natural processes, not power over them or control of them. Wild design is a conversation across time. We listen, ask, and contribute. We inscribe human stories on the larger stories of the system. Participating means living in the systems. We can reciprocate by giving our bodies back to the system. It cannot hurt also to give our minds to the shape of the system. It is knowing what not to do, as well as what to do, when to do it and where, if we do it. The future is already connected with the past through the present. It just gets complex and unpredictable away from the present.

Wild design has to be heroic, especially due to the scale of working on an ecosystem, region or planet. Wild ideas are needed for monitoring natural systems, for closing local loops in energy or matter, for working in closely linked webs, and for initiating connections and collaborations. Our cultures, made more intense in cities and by technology like the web, can be the incubators of new forms of wild design.

Sponsor New Economies from Biodiversity. A nation can discover new economies from biodiversity, such as new crops or pharmaceutical plants. The tax use of such things related to biodiversity could generate income for the nation, as well as limit and protect the wild sources.

Specialize in Trade Specialties. In the past, specialization has allowed human groups to successfully fit the requirements of their environment. Successful cultivation, for instance, intensified the trading of cultivars and resources between groups and permitted further specialization. Village specialization may have been a great adaptation. As a result of surplus food and larger population, specialists can create a flow of specialized objects.

Complex societies depend on production from resources. Increased complexity requires more information processing and more integration of disparate parts. The costs of communication increase. Complex societies need control and specialization. Yet, investment in complexity yields declining marginal returns because of the increasing size of bureaucracies, increasing taxation, and costs of internal control.

Overspecialization reduces flexibility and ability to change, but underspecialization reduces efficiency. Specialization is a way of limiting problems. If the numbers of specialties were reduced, there would be less overhead and higher returns. According to economist David Ricardo, the patchy distribution of resources is not the only reason trade is profitable — trading allows peoples to produce limited items more efficiently, allowing a better payback on their efforts.

Through specialization, now, each nation could become an expert in one area and could reduce taxes in that area. In a global economy, economies of scale are not as important nationally (and therefore, nations do not have to be large at all to have scale advantages, and therefore ethnic groups could form their own nations without giving up some economic advantage). The higher standards of living can be gotten by clever trade and specialization, after self-reliance is achieved.

Balance Total Budgets for a Nation

Nations have traditionally promised security and betterment in return for loyalty and service, but the costs of making good on the promises has forced many nations to increase revenues or reduce their obligations. Reconsider. A nation has to first identify the components, all the components, of the budget, that is, all incoming and outgoing energy and materials, as well as symbolic wealth. Then balance the flows over the desired survival time.

When countries started taxing to raise revenue, it made sense to tax what they could. But, after centuries of dramatic growth and development, these old forms of income need to be shifted to forms that can shape self-reliant and constructive behavior. Perhaps tax shifts can put the budgets of nations back in balance.

Income for a Nation

Income is the form of usable or tradable wealth that is used by the nation to meet its goals and pay its bills. There are many ways of raising funds. A nation could rent its land, sell the services of its citizens, or ask for donations from its citizens. It could sell its resources. Or, it could tax the success of its production and invention.

Taxes for a Nation

This term tax, in its most extended sense, includes all contributions imposed by the government upon individuals for the service of the state, by whatever name they are called: Tribute, tithe, talliage, impost, duty, gabel, custom, subsidy, aid, supply, excise, or other name. A tax is any charge of money or property that imposed by a government upon individuals or entities that are within the government's authority to assess such charges. This term, however, generally does not include charges imposed in exchange for the provision of specific goods or services, such as bridge tolls or sanitation fees.

Although most modern taxes are levied on the basis of economic measurements such as income, consumption, property, and wealth, some governments also impose excise taxes or other taxes. Taxes are usually divided into two great classes, those which are direct, and those which are indirect. The former include taxes on land or real property, and the latter taxes on articles of consumption.

Regarding national-level taxes in the U.S., for instance, the 8th section of Article I, of the Constitution provides that "Congress shall have power to lay and collect taxes, duties, imposts, and excises, to pay" its costs, but "all duties, imposts and excises shall be uniform throughout the United States."

Taxes can have effects other than raising money for a national government. Tax can used as a tool for the equalization of use; many people are luckier or greedier than others. Taxes could be a way of internalizing costs. Taxes can encourage or discourage activities. Today our tax system tends to tax and therefore raise the price of those activities we want to encourage, like investment, employment and property ownership, and tends not to tax or tax at very modest rates those activities that we want to discourage, like pollution and resource exhaustion.

The World Resources Institute argues that current taxes on capital and labor undermine economic efficiency. A tax on capital raises the cost of capital and thus discriminates against technological innovation. A tax on labor raises the cost of labor and thus reduces employment. Displacing these taxes with use and loss taxes would improve the productivity of the economy. For instance, Dower and Repetto suggest that: "Unlike many other sources of federal revenue, a carbon tax would generate overall economic efficiency gains, regardless of how the revenues from the tax are used." A high carbon tax coupled with reduced tax rates on income and profits, according to the World Resources Institute, could generate large gains, not only from improved economic efficiency but from reduced investment in infrastructure and in reduced operating costs due to higher energy efficiency and in reduced environmental damage.

Several European countries, such as Denmark and The Netherlands, are discussing how to achieve the greatest economic efficiency and equity from tax shifting. Sophisticated models have shown that reducing the personal income tax might stimulate short term spending but has modest long term benefits. A better result might occur by directing the environmental tax revenues to expanding investment tax credits. However, that lowers the cost of new capital investments relative to labor and thus could increase unemployment. Also, higher growth could actually increase carbon dioxide emissions. A reduction in payroll taxes could help labor and an increase in investments in energy efficiency and renewables could reduce the linkages between growth and pollution. To balance these changes will be a challenge.

The concept of replacing taxes from production to the environment is a promising shift. It would radically shift the targets of taxation into a more ecological direction. Ecological tax shifts could work in several ways, but all ways would probably have similar key elements. Revenue neutrality would mean that all or virtually all the money raised by increasing taxes on pollution is returned to a community of workers and consumers by lowering or eliminating other taxes on income, payroll, property, or other things. These taxes would not add to the total tax burden, and would even contribute to tax reduction. Another element is a partial exemption for energy-intensive industries to keep them competitive.

An ecological tax shift would not harm the economy of a nation. At the same time it could guide economic behavior towards a more sustainable pattern. For a community, the shift might be transparent or trivial. A similar amount of money given to the government would be returned to the community in services or assistance. For individuals, an ecological tax shift would allow them to have more control over the amount of taxes they pay. In the past, one could only avoid income, payroll or property taxes by going broke, losing a job, or living in a cardboard box. But, one can avoid or reduce ecological taxes by reducing resource use or pollution.

Ecological tax policies could be pro-business, but they would discourage the accumulation of corporate or individual wealth in great quantities. Assuming that the desire for wealth is more related to status than to the sickness of "misplaced concreteness," this should not be undesirable to most wealthy people — they would still have the status of having and displaying more than others.

At the same time as the shift, most exemptions, deductions and loopholes, especially those that are environmentally or fiscally damaging, would be eliminated; these things would include provisions for welfare for mature industries such as oil, mining, timber and automobiles. Government handouts to corporations work against common-sense notions of free markets, innovation and fiscal responsibility. The handouts, billions of dollars per year in the U.S., are paid for by taxpayers. Worse, they contribute to mounting environmental and health costs that US society and the environment must bear.

Use Taxes for a Nation
A use tax for a nation would take a percentage for a service for any resource. It would be applied to those resources held at the national level, although the nation may impose an additional tax on community resources, for the purpose of balancing communities (see elsewhere). A use tax would have the effect of limiting the use of nonrenewable resources, such as coal or oil, as well as the use of slowly renewable resources, such as forests. The rate of the tax would be related to the scale of the economy, as well as to the carrying capacity of the ecological support system.

Air Use Taxes for a Nation. The misuse of air at the local level causes local pollution. Due to its nature, however, air is also a regional and global thing. Taxing air at a national level reflects its regional aspect. Indirect instruments, such as a tax on air, are designed to reduce the scale of output, as important complementary measures in a program of cost-effective pollution control.

Water Use Taxes for a Nation. Water, like air, is part of local, regional and global cycles. Water is used in almost every operation of an industry or bureaucracy. Although people can use water from precipitation, rivers and ponds, much water use comes from aquifers that underlie communities and provinces. This kind of water use has to be taxed at a national level.

Land Use Taxes for a Nation. Where land use is large-scale and very long-term, as in agriculture or forestry, some of the taxes would be national. Agricultural land should also be taxed on its unimproved value, according to Daly and Cobb. National taxes would be collected to pay for regional planning and restoration.

Element Use Taxes for a Nation. Element use taxes for a nation would be applied to every element used for economic purposes, especially those elements which are present in air and water pollution. Many of these elements, especially minerals, can only be used once, although many of them, such as aluminum and iron, can be recycled. Since many of these assets, such as coal and oil, come from federal lands, which are owned by all the citizens of the nation, citizens are entitled to a fair rate of return on the assets. An element tax could capture the value of these assets for current and future generations of citizens.

Carbon Use Tax for a Nation. Carbon is a component of the earth's crust and the atmosphere. In combination with oxygen, carbon forms carbon dioxide, which is recognized as greenhouse gas and has been linked to the greenhouse effect and global warming. William Nordhaus suggested a growing carbon tax to internalize the externality of damage to the climate. This would increase the price of fuels in proportion to how much carbon was released.

The Danish government recently examined the impacts on their economy if they were to increase their carbon dioxide tax by about $25 per ton. The conclusion was that if the revenue generated were returned through income tax reductions, there would be a loss in production and a rise in unemployment, but if the revenue generated were returned by reducing the social security obligations of a business, employment and production would both rise.

Nitrogen Use Tax for a Nation. Nitrogen is also a critical element of life; it is a major part of the atmosphere. Nitrogen oxides play an important role in the atmospheric reactions that generate ground-level ozone, which contributes to smog, and acid rain. The U.S. Environmental Protection Agency (EPA) believes that nitrogen oxides can irritate the lungs and lower resistance to respiratory infections such as influenza. NO_x and pollutants formed from it can be transported over long distances, so problems associated with the pollutant are not confined to areas where it is emitted.

Nitrogen gas is emitted into the atmosphere by cars and industrial processes. Some of it then returns the planet in rain, which can harm plants by changing the nutrient content of soil. Emissions from industrialized nations seem to be stabilizing and nitrogen deposition is even declining in some regions. Rapid population growth and industrialization means developing countries are becoming major emitters of nitrogen.

Since the early 1990s, Sweden has imposed an environmental charge on NO_x emissions from large combustion plants. The fee is $2.18 per pound, measured as NO_2. The revenue from the charge is returned to the plants in proportion to their energy production. This refund policy allowed the tax to gain business support, yet the refund policy encourages NO_x reductions.

Sulfur Use Tax for a Nation. Sulfur dioxide and NO_x emissions are considered the main contributors to acid rain, which the U.S. EPA believes degrades surface waters, damages forests and crops, and accelerates the corrosion of buildings. The Clean Air Act Amendments of 1990 adopted a program to control acid rain; it introduced a market-based system for emission allowances to reduce SO_2 emissions. An emission allowance is a limited authorization to emit a ton of SO_2. The EPA allots tradable allowances to affected electric utilities according to their past fuel use and statutory limits on emissions. Once the allowances are allotted, the act requires that annual SO_2 emissions not exceed the number of allowances held by each utility plant. Firms may trade allowances, bank them for future use, or purchase them through periodic auctions held by the EPA. Firms with relatively low abatement costs have an economic incentive to reduce emissions and sell surplus allowances to firms that have relatively high abatement costs.

Another option is to tax emissions of SO_2 from stationary sources not already covered under the acid rain program. Imposing a tax of $210 per ton of SO_2 emissions from those sources would raise more than $6 billion over the 2001-2010 period, according to researchers. Most firms do not pay taxes or fees on emissions that regulations still allow, although major stationary sources must pay fees annually to cover program costs of operation permits under the U.S. Clean Air Act Amendments of 1990. Basing the tax on the terms granted in those air pollution permits would minimize the cost of administering the tax for the Internal Revenue Service. Opponents to the tax argue that it would impose an additional burden on many firms that already incur costs to comply with current regulations on emissions.

There are market incentives and disincentives. Sometimes marketplace strategies have been coupled with regulatory strategies. For example, the nation could impose a cap on sulfur emissions from power plants and then create an emission offset trading system to lower the overall cost of sulfur reductions. One might argue that if the tax is stiff enough, there is less reason for an accompanying prohibition. Sweden's tax on sulfur emissions is about four times the cost of sulfur reduction. As a result, after the imposition of the tax, sulfur emissions fell by 40 percent from oil refineries.

Other Element Use Taxes for a Nation. There is a U.S. Federal Tax on Ozone Depleting Chemicals. In 1989 the U.S. Congress enacted a tax on eight ozone depleting chemicals as part of its Omnibus Budget Reconciliation Act. It extended this tax to 12 additional chemicals and raised the tax on the original 8 chemicals in the National Energy Policy Act of 1992. The Clean Air Act established caps on most chlorofluorocarbons (CFCs), with a phase out occurring around the year 2000. The tax on CFCs was $1.37 a pound in 1990 and 1991, about twice the then current product price. Recycled CFCs were exempted from the tax. The tax was raised in 1990 and again in 1992. The tax rises to $3.10 per pound in 1995 and then rises by 45 cents per pound per year thereafter. The tax is proportional to the chemical's potential for depleting the ozone layer.

In another instance, ozone at ground-level has remained a pervasive pollution problem in many areas of the United States. To control ozone pollution, the EPA has traditionally focused on reducing emissions of VOCs (and, more recently, NO_x). VOCs include chemicals such as benzene, toluene, methylene chloride, and methyl chloroform. VOCs are released from burning fuel, from fossil fuels or wood, or from using solvents, paints, glues, and other products. The EPA can tax emissions of VOCs from stationary sources, such as industrial facilities, such as chemical plants, petroleum refineries, and coke ovens, to small sources, such as bakeries and dry cleaners. The vast number and diversity of stationary sources make it difficult to estimate emissions and the cost of abatement. A tax of $2,300 per ton on all VOC emissions from stationary sources might promote some abatement and could generate slightly over $111 billion in revenues from 2001 through 2010. It could apply to mobile sources.

Species Use Taxes for a Nation. Tax would be charged on the taking of members of any species. Many species are slowly renewable or functionally nonrenewable. Many species move between local human communities. This national tax would work to encourage preservation, take-limits, or take-efficiency at the national level.

Loss Taxes for a Nation
A loss tax is a tax on losses from the capital base, that is, it is a tax on the destruction of resources, not just their use or on their negative impacts. It is similar to a dispossession tax or a capital depletion tax. For a nation, it is a tax on losses from the national base. Functionally nonrenewable resources are often distributed across local boundaries, and should be managed at a national level.

This kind of loss tax is on things or processes that interfere with other things and processes, things that cause runaway feedback or the destruction of cycles, things in other words that reduce our continued use of and enjoyment of the earth. Many of these things have been subsidized for many decades as a result of the power of special interests. The purpose of this kind of tax is to change behavior that depletes resources and discourages labor. This tax would have the effect of internalizing both ecological and social costs; since all consumers would be paying the real costs, no consumers would be protected.

Land Conversion Loss Taxes for a Nation. This tax would be on the conversion of complex systems to simpler, and more expensive to manage, systems, for instance, the conversion of forests to agricultural fields, or the conversion of fields to asphalt-covered parking lots. This tax on a national level would pay for the required planning to keep land in optimum national or human coverage.

Nonrenewables Loss Taxes for a Nation. Nonrenewables includes geothermal energy, as well as fossil fuels, such as oil and coal. These resources represent uniquely rich resources and have to be taxed at a national level due to their uniqueness. Furthermore, they should only be used as transitional resources, although they have been treated as limitless resources in the past.

Geothermal Loss Taxes for a Nation. Geothermal energy would be taxed because it is not permanent; it is the long-time flow of the energy of formation of the planet to space. The U.S. Geothermal Energy Act of 2004 (H.R. 4094) would reduce that nation's reliance on oil, gas, and coal-fired electrical power plants by authorizing a new assessment of that nation's geothermal resources and expanding geothermal energy investment by offering tax incentives to develop these resources. However, the use of that nonrenewable source would be taxed at a very low rate (if at all).

Fossil Fuel Oil Loss Taxes for a Nation. Oil and coal are often located under national territories and may extend between borders. Nations should tax them due to their nonrenewability. And, they should be used as transitional energy sources, now that their maximum economic extents have been basically mapped and their economic cost calculated.
   Oil and gas should also be taxed because their production and use can damage and destabilize ecosystems. Oil use places extra carbon and pollutants in the air. A high tax on oil would reduce its use, as well as encourage renewable, cheaper alternatives. An example extraction tax would be 50 percent of the cost of a barrel.

Natural Gas Loss Taxes for a Nation. Because of its distribution, natural gas may be taxed more effectively at a national level. This would allow better planning and transition to renewable fuels.

Coal Loss Taxes for a Nation. Many environmental and medical costs would have to be addressed at the national level. This additional tax would pay those costs, as well as additional costs for planning at the regional level.

Slow Renewables Loss Taxes for a Nation. These resources, such as forests and fisheries, are essentially nonrenewable in terms of a human lifetime. They are also not distributed uniformly and cross community boundaries.

Forests & Trees Loss Taxes for a Nation. Nations need to tax the losses of forests and trees, especially considering the status of the earth as a forest planet. Forest land should be taxed on its land value. Since the forest is the capital, and trees characterize the forest, the depletion tax on trees needs to be high. If enough trees are removed the forest dies and the capital can be lost. This tax might be specific to national lands, forests and parks.

Fisheries & Fish Loss Taxes for a Nation. Fish can be caught without being produced. The constancy of the ocean, lake, river, and stream environments, even with tides and shifts, has allowed many kinds of animals to live in large populations. Populations can shift to different norms with exploitation. Some populations are stable, but not resilient, such as U.S. Great Lakes fish. C.S. Holling notes the pattern of fishing pressure in the history of the Great Lakes: There is a period of intense exploitation, during which there was a prolonged high-level harvest, followed by sudden and precipitous drops in population. Although not entirely unexpected for sturgeon, with slow growth and late maturity, it was unexpected for herring and whitefish. Fishing pressure shifts the age structure of populations toward younger ages. Apparently, fishing progressively reduced the resilience of the system, so that when an inevitable unexpected event took place, the populations collapsed, with a ripple effect on fish-dependent species, including birds and parasites.

Water tends to cross ecosystem boundaries; fish also move between ecosystems, quite dramatically in the case of salmon. A national tax on fisheries (the fish-bearing environments) and species takes could allow some populations to recover.

Fast Renewables Loss Taxes for a Nation. Fast renewables are things that are usually part of daily, monthly or annual cycles, such as solar power, hydropower or wind power. They can be taxed at a relatively low rate or even a negative rate to encourage development.

Wind Power Loss Tax. Wind power would have a negative tax (or tax credit) of 1 cent per kilowatt-hour for electricity produced from or for national interests.

Solar Power Loss Tax. Solar power would have a negative tax of 2 cents per kilowatt-hour for electricity produced from or for national interests. There would be tax credits for solar domestic or cross-community business water heating.

Hydropower Loss Tax. Hydropower would have a negative tax of 1 cent per kilowatt-hour for electricity produced from or for national interests.

Waste Loss Taxes for a Nation. Various wastes, from solid wastes to water and energy, would be taxed at amounts necessary to discourage their use or encourage their incorporation in industrial cycles. A national tax on waste is an additional tax designed to catch oversights or adjust equalizations in a process.

Adjustment Taxes for a Nation
Adjustment taxes are regulation or correction taxes to regulate or harmonize cures and causes. Many forms of liquid or solid waste would be taxed by

volume, for example, in the U.S. at a rate of $0.50 per cubic foot. Other forms of solid waste, such as hazardous waste, tires, batteries, or nuclear by-products, would be taxed sufficiently to pay for their isolation or breakdown. Recyclable solid waste, such as animal excrement or rock, would not be taxed if it was recycled properly. A national tax would allow the nation to recover specific costs for special wastes that a community might not be able to contain.

Sin Adjustment Taxes for a Nation. Human sin tax is collected to offset the consumption of activities that would require eventual health care. If people do dangerous things, and expect their government to take care of them, then government has to tax those things that result in illness. This tax would benefit people who suffer from bad habits and offset their health care. Due to its relatively high rate, it might reduce the consumption of addictive substances. The national tax would be used to coordinate and equalize health care costs.

Cigarette Adjustment Taxes for a Nation. Inhaling the smoke from burning in plants in moderation may be stimulating, especially in ceremonial settings. However, overuse has serious health consequences. An industry exists that makes profits by selling such things to users and addicts. Like an excise tax, an adjustment tax on sales of cigarettes is a fixed fee on each pack of cigarettes sold. The cigarette tax for the nation would level out community taxes. It varies by state in the U.S. and ranges from $0.07 per pack in South Carolina to $2.46 per pack in Rhode Island. The tax doubles or even triples the retail cost of cigarettes in some states.

Alcohol Adjustment Taxes for a Nation. This relatively high tax would offset the increased medical and social expenses from its use. Taxes at a national level would fill community gaps and ensure coverage for those harmed by the personal and social effects of excessive use.

Drug Adjustment Taxes for a Nation. Many nations try to control the kinds of drugs used. Legal drugs would be taxed according to their purpose or perceived social benefit. Illegal taxes would be taxed at a higher rate. If all substances were legalized, then taxes could be used to control their use. There could be additional social benefits, such as decriminalization and deimprisonment, and all the costs and savings that entails, again.

Pollution Adjustment Taxes for a Nation
All forms of pollution would be taxed, especially those related to regional and global cycles of the elements necessary for metabolism of living beings.

Another way to reform the tax code is to introduce pollution charges on the amount of pollution that a firm or product releases into the air, water, or soil. A pollution tax would be a means of tackling the 'market failures' that arise when businesses and consumers are not confronted with, or responsible for, the full health and environmental costs associated with their activities. Pollution taxes make polluters pay for their damages and incorporate these costs into their decisions and product prices.

There is no doubt that introducing new pollution charges would be challenging in any political system. As a result of declining income and growing budget deficits, however, this approach may become more attractive. Revenue from pollution charges could be utilized to dispense with distortionary taxes as part of an innovative, revenue-neutral tax reform. For instance, they could offset payroll or income taxes.

Pollution charges should not even involve additional disruption; they are being used in many developed countries and several U.S. states. China is using charges to address some of its environmental problems, including water pollution.

Industrial Pollution Adjustment Taxes for a Nation. As a direct result of many industrial processes, solid particles and liquid droplets can be found in the air, ground or water, in a variety of sizes and concentrations. According to U.S. Environmental Protection Agency (EPA) studies, the emissions of particles, especially if combined with other air pollutants, are linked to some adverse health effects. For example, particulate matter can carry heavy metals and cancer-causing organic compounds into the lungs, increasing the incidence and severity of respiratory diseases. Other health effects may include chronic bronchitis.

Since monitoring systems and a permit system are already in place for coarse particle emissions, emissions from stationary sources could be taxed. That tax could be administered similarly to the taxes on sulfur and nitrogen oxides. A relatively small tax per ton of coarse particulate matter could raise over half a billion dollars per year. The national tax on would address mobile pollutions and even out disparities at the community level.

Carbon Pollution Adjustment Taxes for a Nation. A carbon pollution tax would link the effects of burning these fuels to the cost of repairing the damages from burning. A carbon tax could greatly offset other tax rates, as well as reduce energy consumption, especially fossil fuel consumption, and address climate change. A number of studies have evaluated the macroeconomic impact of a tax on carbon equal to $100 (USD) a ton. Although there would probably be a small reduction in GNP, which is itself a flawed measure, there would be offsetting tax reductions elsewhere in the economy; and, some of the tax money would go toward improving efficiency.

Displacing these taxes with carbon and other pollution taxes would improve the productivity of the economy. A high carbon tax, coupled with reduced tax rates on income and profits, according to the World Resources Institute, could generate a possible gain of 45-80 cents per one dollar of tax shifted. The gain would come not only from improved economic efficiency but from reduced investment in infrastructure, in reduced operating costs due to higher energy efficiency, and in reduced environmental damage.

Phosphates Pollution Adjustment Taxes. Phosphorus is an element. Phosphate is a salt of phosphoric acid. The tax is for the mineral phosphorus content, which is used in animal feed and can cause pollution; phosphorus may also be related to red tide. A tax might reduce excess phosphorous used in agriculture by limiting the use of phosphate in animal feed. This would require knowing what stock owners have, and letting them know how much and when to pay.

Nitrogen Pollution Adjustment Taxes for a Nation. Nitrogen is a significant part of the earth's atmosphere. Nitrogen gas emitted into the atmosphere by cars and industrial processes can be a pollutant, however. Some of it then returns the planet in rain, which can harm plants by changing the nutrient content of soil.

Researchers have warned that rising nitrogen emissions from developing nations will soon threaten plant life in some of the most biodiverse parts of the planet. A team led by Gareth Phoenix of the University of Sheffield has shown that, in the mid-1990s, the average amount of nitrogen deposited on the planet's 34 biodiversity 'hotspots' was more than 50 per cent higher than the global average. They say this figure could more than double by 2050, at which time nitrogen levels in 17 of the 34 hotspots will exceed critical levels that European nations have set to protect their sensitive ecosystems. As a result, many of the hotspots will soon be in danger of being damaged by high levels of nitrogen, say the researchers. They add that this may already be true for some areas. The researchers point out that rapid population growth and industrialization means developing countries are becoming major emitters of nitrogen. Emissions from industrialized nations, on the other hand, are stabilizing and nitrogen deposition is even declining in some regions. This suggests developing countries are responsible for the damage suffered by biodiversity hotspots.

Since January 1, 1992 Sweden has imposed an environmental charge on NO_x emissions from large combustion plants. The fee is $2.18 per pound, measured as NO_2. The revenue from the charge is returned to the plants in proportion to their energy production. This refund policy allowed the tax to gain business support, yet the refund policy still encourages NO_x reductions. The average cost of reducing one kilogram of NO_x is $1.20 while the tax is $2.18 per pound.

Sulfur Pollution Adjustment Taxes for a Nation. Sulfur is an element necessary for life. It is also used for processes in refineries, smelters, mills, and chemical plants, often producing sulfur oxides as pollutants that contribute to smog and respiratory diseases.

Some nations impose a cap on sulfur emissions from power plants and a tax on sulfur emissions. Sweden's tax on sulfur emissions is about four times the cost of sulfur reduction. As a result, after the imposition of the tax, sulfur emissions fell by 40 percent from oil refineries.

Elements & Compounds Pollution Adjustment Taxes. In 1989 the U.S. Congress enacted a tax on eight ozone depleting chemicals as part of its Omnibus Budget Reconciliation Act. It extended this tax to 12 additional chemicals and raised the tax on the original 8 chemicals in the National Energy Policy Act of 1992.

The Clean Air Act established caps on most chlorofluorocarbons (CFCs), with a phase-out occurring around the year 2000. The tax on CFCs was $1.37 a pound in 1990 and 1991, about twice the then current product price. Recycled CFCs were exempted from the tax. The tax was raised in 1990 and again in 1992. The tax went to $3.10 per pound in 1995 and then rose by 45 cents per pound per year thereafter. The tax is proportional to the chemical's potential for depleting the ozone layer.

Many other elements and compounds are toxic, and are used in cleaners and biocides. Taxing them at a national level would coordinate use and research, as well as limits.

Agricultural Pollution Adjustment Taxes. The use of fertilizers and pesticides, as well as specialty and energy-intensive machinery would be taxed at a national level to offset their costs and effects.

Pesticides Adjustment Taxes for a Nation. Because of the adverse impacts on human health and the environment, as a result of pesticide use, and the direct financial costs, such as the treatment of water, and wider environmental costs, such as loss of biodiversity, which are much harder to identify and value, there would be national taxes on pesticides (perhaps under an umbrella term, like biocides).

Fertilizer Adjustment Taxes for a Nation. Fertilizer is used to increase the productivity of crops, but it causes problems. Even natural fertilizers, such as dung or plant matter, can cause pollution and damages at large scales. The application of national environmental tax shifts to pesticide and fertilizer use could provide an incentive to minimize pesticide and chemical fertilizer use and could generate revenue for tax relief in other areas.

Water Adjustment Taxes for a Nation. The pollution of water from water courses or aquifers would be taxed by a national government. The rate would be equal to or greater than the cost of cleaning or purifying the water.

Personal Pollution Taxes for a Nation. A personal pollution tax would be imposed on the use of things that create pollution through personal use, such as packaging or fuel. A fuel tax, for instance, would be imposed on the sale of all fuel regardless of use.

Heritage Items Sale Adjustment Taxes for a Nation. A heritage tax would be applied to any thing considered to be part of a natural or cultural heritage, such as natural bridges or famous bridges. It would include unique art works produced by people in a culture. It would also include a tax on the export of anything without a value-added component, such a raw logs or raw minerals.

The Heritage tax would be similar to a tariff. One function of a tariff would be to protect special accomplishments or special resources; another would be to raise money for the government and to protect home industries. Too high a tariff would prohibit all imports. Too low a tariff would not protect home industries from those who had less environmental protection or paid fewer employee benefits. Tariffs could be used to encourage economic self-sufficiency and give a government more control of its economy.

Financial Speculation Adjustment Taxes. The speed and detachment of money as a medium of exchange may artificially change the values of things that should remain uninflated or undepressed. A speculation tax, like a Tobin tax, is a small tax on each international financial transaction.

There are many potential benefits from a modest tax on financial transactions, such as the buying and selling of shares of stock or blocks of foreign currencies. Such a tax would have the effect of reducing short-term speculation in these markets, thereby making them somewhat less volatile. It would slow capital movements. Computers have given traders incredible advantages at unimaginable speeds. The tax would also cut back some of the economic resources that are wasted in these transactions, since if the number of trades declines, the money spent on these trades would decline as well. In addition, it would make the tax code fairer, since most financial speculation is conducted either directly or indirectly by wealthy people.

But, it would have to be collected in every country. Otherwise some countries would not collect it to increase their advantages. One or two countries could not afford to impose such a tax, unless all countries were to cooperate. If one country imposed a speculation tax unilaterally, then traders would simply move their business to a nation that did not tax their transactions; the internet, for instance, has given traders an unprecedented mobility. Unless mobility were controlled, the main effect of the tax would be to deprive the markets of the pioneer country of financial business. If large trading blocs were to adopt it, like the European Union, and it was shown to be manageable, then an international agreement on speculation tax, binding on all countries, could be arranged at the global level.

As with all taxes, there will be opportunities for evasion. The incentives for evasion are much greater in the case of money laundering, but there are some effective regulations on money laundering. The demands on financial institutions under those regulations are comparable to the demands that would be imposed with the imposition of speculation taxes. The incentives for evasion in case of copyright laws is also great. The fact that governments can protect copyrights for publications and internet suggests that a speculation tax could be enforceable. Collection of the tax at the national level would help with coordination and enforcement.

Distribution Taxes for a Nation
Distribution taxes are the same as reapportionment for luxuries and big incomes. Combined with heroic inheritance and profits, these incomes concentrate wealth and essentially remove their receivers from any sense of participation in the community or nation. Although such taxes would take more from those who have more, they should not discourage people from working to have more, as well as getting higher status from paying more.

Heroic Possessions Distribution Taxes for a Nation. A heroic possessions tax is a tax on products that are not considered essential, in other words, luxuries. A luxury tax is similar to a sales tax or VAT, except that it mainly affects the wealthy because the wealthy are the most likely to spend heavily on luxuries such as expensive cars or jewelry.

The provisions of a luxury excise tax in the U.S. were contained in the Omnibus Budget Reconciliation Act of 1990, which was intended to reduce the federal deficit by enforcing a 10 percent surcharge on high-priced products of the automobile, boat, aircraft, jewelry, and fur industries. Effective on January 1, 1991, the tax applied to the 'first retail sale' of luxury goods with a sales price above the following thresholds: automobiles $500,000; boats $100,000; aircraft $25,000; and jewelry and furs exceeding $10,000. The tax was projected to provide revenues of $9 billion in the first five years. It did not.

There was controversy from ambiguities in the determination of tax liabilities. There was concern that the tax was unjustly levied on only those five industries. There were difficulties in the administration of the rule in terms of compliance of the five industries. There was criticism of subjective loopholes of the regulations, evident in the cases of the boat and automobile industry. It was argued that the imposition of the tax aggravated the plight of luxury industries already depressed by current economic conditions and losing jobs and profits. Legislators levied the tax on only five industries, to make the tax easier to administer. Electronics was not used a category because of different types of dealers and equipment. Houses and furnishings were not considered.

For the consumer, the tax seemed too complex and the thresholds difficult to understand or apply. For opponents, the tax was worrisome because, once the statute was in place, it would be simple to lower the threshold amounts or to increase the rate. Whereas autos, planes, and boats require licenses and leave a paper trail, jewelry and furs can be purchased and hidden, making it difficult to collect a tax.

Although a luxury tax of some sort may be justified on many things, the regulations have to clarify implementation in order to achieve at least some degree of equity. This tax would be coordinated at the national level, considering the multiple residences of those who can afford other luxuries. Regulations have to broaden the definition of luxury. It is hard to argue that a luxury tax is unjust and counter-productive, considering the massive suffering and starvation outside the gated compounds.

Heroic Income Distribution Taxes for a Nation. A heroic tax would be applied to income over an acceptable ratio. Herman Daly and John Cobb quote a range of the acceptable inequality of income at ten to one, although some corporations, like Ben & Jerry's ice cream used to limit it to a one to seven ratio. As J.B. Cobb and H. Daly point out, the idea of unlimited inequality works against the notion of community. As they also wisely point out the goal of a community is not some perfect ideal of equality, but a limited inequality that allows individual differences to show, and individual rewards for luck or skill, but also allows other individuals to catch up within a decade or generation.

A heroic income tax would not prevent some people from getting richer, but it would slow the rate and limit the vast differences. The poverty line in the U.S. in 2005 for one person is $9570; for a family of four, it is $19,350 USD.[12] There is no maximum income, although surprisingly many people receive many millions per year. That would mean that if the minimum for a family of four in the United States was set to $20,000, the maximum income for a family would be $200,000, at a ratio of one to ten. Although limiting wealth to a ratio might seem like unfair and confiscatory taxation, it would not be dissimilar to the way that many people achieved great wealth — by confiscating free goods of nature or cheating the less powerful. So, this tax rate would be quite high for those with heroic incomes. Paying such a tax might make those people heroes.

Heroic Transfer Distribution Taxes for a Nation. A transfer tax is a tax on any transfer of wealth, including inheritance, death duty, estate or bequest. It is applied to any transfer of wealth, to spouses, children, relatives, and even bequests to trusts, foundations and charities, and not on the total wealth.

Due to a unique set of circumstances in the past, including luck, theft, hard work, and inheritance, many large fortunes do not represent taxed income or savings, but they do represent unrealized capital gains which would never be taxed as capital gains under the U.S. federal income tax, for example. Those most affected would be estates of considerable size, after applicable credits. The effect on family-owned farms would be minimal. The American Farm Bureau Federation acknowledged that it could not cite a single example of a farm having to be sold to pay estate taxes.

This tax would require heavier regulation. This tax would serve to prevent the perpetuation of wealth, free of tax, in wealthy families. It would work towards reducing the accumulation between generations, thus leveling the field more for the next generation.

Heroic Profit Distribution Taxes for a Nation. Two words: ExxonMobil. Tax heroic profits from rare, essentially nonrenewable resources to pay for alternative sources and technologies.

Discussion of Missing or Transitional Taxes for a Nation

Under this scheme, there would be no national taxes on buildings, equipment or inventories. There would be no national corporate income tax, although, as Daly and Cobb suggest profits would have to be distributed to all shareholders as income. There would be no personal income tax, no property tax, and no sales tax, so that things people want or work for should be encouraged. Once the real price of oil and other resources has settled out, no other taxes are needed for value-added efforts or goods. In order to reach this position, however, there may be a series of transitional taxes, such as an added tax on all new vehicles.

License Privileges for a Nation

A license is a formal authorization by law to do something, such as marry, hunt, or practice law or medicine. The granting of licenses has a long history, in many nations and cultures.

Marriage License for a Nation. A marriage license is permission from a legal authority for the marriage of two people to be performed. The requirements differ depending on the time and place. Combined with other rights or privileges, such as having children, a marriage license would be required to indicate preparedness and intent. Although this license is required at a community level, it would be available at a national level, also.

Reproduction License for a Nation. In the early 1970s, former U.S. Senator Bob Packwood and the late Senator Jacob Javits seriously discussed having the government license parents to have children. A reproduction license would be permission from a legal authority for two people to have offspring. The requirements would differ depending on the time and place. The purposes of such a license would be several, starting with qualifying the parents, through a reading program or knowledge of health issues, such as smoking or drinking, and continuing with the health and safety of children, and with consideration of the effects on the health and fitness of the entire gene pool and on society.

Although this license would be granted at a community level, it might be coordinated or duplicated at a national level. The penalty for not having the license might range from a fine to reallocation of the child to another family or institution, although that alternative would have to have numerous safeguards for the health and safety of the child.

One incentive for having the license might be a monetary rebate. Another incentive could be an extra coupon or grant towards education or luxuries. This license would not limit the number of children or help with the planning of institutions, such as schools—that would be the purpose of tradable vouchers to have children.

Driving License for a Nation. A driver's license is an official document which states that a person has the necessary qualifications to operate a motorized vehicle, such as a motorcycle, car, truck, camper, trailer, or a bus. Driver's licenses are generally issued after the recipient has passed a successful driving test and proven that they meet the age, education and understanding requirements. Although licenses would be awarded at a community level, they might be coordinated or duplicated at a national level, especially for national service or representation.

Voting License for a Nation. A license to vote would be required to demonstrate that the voter is living, lives in the community, speaks the language, and understands the basic issues. This could prevent some kinds of fraud. Although this licenses would be awarded at a community level, it would be duplicated at a national level for national elections.

Media Licenses for a Nation. Any use, for broadcast or communication, of common frequencies of the spectrum of waves, would be licensed by the agency responsible for regulating interstate and international communications by radio, television, wire, satellite, or cable. This license would be coordinated with community-issued licenses.

Business License for a Nation. The nation would only need to issue a separate business license if the business operated across communities, provinces or states.

Collecting License for a Nation. Anyone who collects plants, animals, fish, or any living substance for its compounds or DNA, would be required to have a national license for collecting in national parks or preserves. The license would certify that they had the knowledge to collect, and it would limit the number of people collecting. Issuing licenses would help protect national commons, such as parks and wilderness areas, as well as coordinate the timing and limits of collection. The nation would only need to issue a separate license if the business operated across provinces or states or on federal lands.

Weapons License for a Nation
A weapons license would verify that knowledge and experience requirements are met and reviews and researches of criminal history have been conducted for information that might preclude the issuance of a license, such as a criminal record.

  The nation would have a bureau responsible for the issuance and denial of licenses. This bureau would receive and examine licensing applications for statutory compliance and verify the applicant's eligibility for licensing through former employers, educational facilities and examination of criminal history records. The kinds of weapons could be limited to knives, arrows, single-shot firearms and multi-shot fire arms; certain weapons, such as automatic guns, tanks, bazookas, and missiles, would not be licensable. Applying for a license to carry a weapon for self-defense is a right of the law-abiding citizens. Appropriate personal weapons can be carried responsibly, properly and safely.

A nation needs to stop the spread of assault rifles, pistols, hand grenades, mortars, and other so-called light weapons. There is a growing realization among governments that in the post-cold-war world that the high-powered automatic small arms, more than cannons or missiles, are the prime contributors to regional instability, fueling ethnic or nationalistic wars with uncountable casualties.

Governments have to decide whether to concentrate on illicit trafficking or to address the legal trade as well. Many governments consider controlling the legal trade crucial to keeping weapons out of regions of conflict and the hands of dictators.

Users Fees & Tolls for a Nation
Fees could be charged by the nation for use of national space or renewal cycles, for example, visiting fees or sanitation fees. Tolls could be charged by the nation to cover extra expenses for transportation, for instance, bridge tolls or road tolls on national highways.

Conservation & Tourism Fees for a Nation. Historically, many areas get a substantial part of the money used to market themselves as a destination from a room tax or surcharge on hotels and motels. This is commonly referred to as a 'tourism tax' because the vast majority of it is paid by tourists, as opposed to local residents, and much of the income is used to promote tourism. Although there are always exceptions to the general rule, typically, if this tax is expanded to other areas, it is generally tourist specific, such as tickets for attractions or for food and souvenirs. Tourism is often a considerable source of income for nations with dramatic scenery and wildlife. These fees are also needed to repair the damage caused by tourists. Tourism may have to be limited to reduce their impacts on other cultures and on the environment.

Cultural Treasures Fees for a Nation. Although much of the heritage of the nation would be free to observe or visit, fees could be charged to limit access. Fees would also be used for special costs such as cleanup.

Labor Use for a Nation
Historically, rulers have exchanged protection and coordination for the labor services of residents or citizens, especially for large public or national projects. Although the use of labor has become much more formal, volunteers are a major source of service.

Require Service to Nation for Two Years. Nations need the services of citizens to carry out many of the plans and operations of the government. A national service provides a pool of young or skilled people who are committed to working for a certain length of time. For the people themselves, service provides experience, wages, and a sense of need. National service has been usually understood to mean a form of military service in which all citizens, or all male citizens, can participate, either voluntarily or involuntarily. The involuntary form of service has been called conscription or draft.

Conscription is a general term for involuntary labor demanded by some established authority, but it is most often used in the specific sense of government policies that require citizens to serve in their armed forces. In the U.S. the conscription program was known colloquially as 'the draft.' Many nations rely on a volunteer or professional military most of the time, although they reserve the possibility of conscription for wartime and 'crises' of supply. Conscription has also sometimes been used as a general term for nonmilitary involuntary labor demanded by some established authority; for example, in Japan during World War II, Japanese women and children were conscripted to work in factories. This kind of service may last for an indefinite time, such as the length of a war or an emergency.

National Service is still used to describe the compulsory military service that is still implemented in some countries, including Singapore, Greece, Germany, South Korea, Israel, Republic of China (Taiwan), Russian Federation. According to the U.S. CIA, comprehensive national service incorporates the Army, Navy and Marine Corps, Air Force, and Coast Guard, and all these branches are controlled by the president. This service usually has a fixed period of time, such as one to two years.

Every nation should require a two-year period of service to the nation, for civil work, police work, or overseas work (as with the U.S. or Switzerland service corps). Every person would have the choice of area of service. The nation would provide training in each area.

Encourage Volunteers for a Nation. Some people may wish to voluntarily extend their conscription for a length of time, depending on the project. Many forms of volunteer work take place in a private setting or for an NGO dealing with social problems or environmental conservation. Thus, people volunteer for hospitals or libraries, food programs or the humane society. A nation could encourage this or support it, through recognition or reward. A nation could also develop its own specific programs, for national parks or political activism.

Discussion of Income for a Nation

Most income would come from taxes on inefficiency and waste, as well as from taxes on heroic profit and ownership. A tax on these activities should not dampen the competitive spirit displayed by many citizens.

Some people worry that these taxes would undermine the competitiveness of domestic industry. This concern has been raised with regard to national green taxes in Europe, where countries operate under the Single Europe Treaty, which forbids taxing imports within Europe, and with regard to state taxes in the U.S., where the Constitution prohibits states from imposing taxes on interstate commerce. Countries and states may prefer to introduce shifted or green taxes cooperatively, rather than unilaterally.

Shifted or green taxes should not impose a competitive disadvantage if imposed on the vernacular economy, on the household sector, or on that portion of the business sector that does not export its products or services. The most significant impact would probably fall on energy-intensive, exporting industries. For a carbon tax, for instance, the greatest burden would fall on a energy-intensive industry, such as Portland cement manufacturers, which is rarely an exporting or interstate industry, because of its high transportation costs relative to its modest value. Nitrogenous fertilizers and a few primary metal industries, such as aluminum and steel, would experience price increases in excess of 0.5 percent, but those industries have other difficulties being competitive in some nations.

Payouts for a Nation

The nation has to pay for its existence, as well as the expenses, amenities, and services that it trades for its existence. In economics, business and accounting, a cost is the value of inputs that have been used up to produce something, and hence are not available for use anymore. In business, the cost may be one of acquisition, in which case the amount of money expended to acquire something is counted as cost. In this case, money is the input that is gone in order to acquire the thing. This acquisition cost may be the sum of the cost of production as incurred by the original producer, with the further costs of transaction incurred by the buyer beyond the price paid to the producer. Usually, the price also includes a mark-up for profit. Costs are often further described based on their timing or their applicability.

When a transaction takes place, it typically involves both private costs and external costs. Its private costs are the costs that the buyer of a good or service pays the seller. Its external costs, also called externalities, are the costs that people other than the buyer are forced to pay as a result of the transaction. The bearers of such costs can be either individuals, communities or the nation (or the ecosystem or planet). External costs are often nonmonetary, and as a result are difficult to quantify for comparison with monetary values. They include things like pollution, things that a nation will probably have to pay for in some way at some time, but that are not included in transaction prices.

Social costs are the sum of private costs and external costs, that is, both the costs internal to the firm's production and external costs not included in the firm's production. For example, the purchase price of a car reflects the private cost experienced by the manufacturer. The air pollution created in the production of the car is an external cost. The manufacturer does not pay for these costs and does not include them in the price of the car, so they are external to the market pricing mechanism. The air pollution from driving the car is also an externality. The driver does not pay for the environmental damage caused by using the car. A psychological cost is a subset of social costs, for example as a result of endangering pedestrians, that specifically represents the costs of stresses or losses to quality of life. A community has to consider all costs for as long as possible into the future.

Reduce Legacy Debt of National Government. Debt reduction, especially in countries like the United States, is necessary to reduce the burden on future generations who did not receive the direct, initial benefit. Unfortunately, this may mean some sacrifice on the part of many who also did not benefit from the incurrence of the debt, although in some cases, it was themselves or their parents who benefited.

Operate National Government. The government itself has expenses for its maintenance and activities. The government has to pay for information services, as well as for collection and administration costs.

Elect Leaders & Representatives at Government Cost. The government has to pay for its leaders and representatives, as well as for staff to provide services. A government could consider aligning salaries to a national mean, and making it illegal to accept any campaign contributions from special interests that would expect to have special favors done for them. Tying benefits, medical and retirement to a national mean would also work to normalize government service related to the public. Only the government would pay for elections. Individual and corporate monies would be forbidden.

Open Communications Paths for Nation. The government has to ensure that there is equal access to media. In the U.S., for instance, it has to allow more fair use of common waves in the national atmosphere.

Fund Special Departments of Nation. Research benefits a nation in many ways. Research funding is a term generally covering any monies for scientific research, in the areas of 'hard' science and technology, social sciences and humanities. The term often connotes funding obtained through a competitive process, in which potential research projects are evaluated and only the most promising receive funding. Such processes, which are run by government, corporations or foundations, allocate scarce funds. Total research funding in the richest countries is between 1.5% and 3% of GDP; Sweden is the only country to exceed 4%.

Most research funding comes from two major sources, corporations, through special research and development departments, and the community or national government, whose work is primarily carried out through universities and specialized government agencies. Some small amounts of scientific research are carried out (or funded) by charitable foundations, especially in relation to developing cures for diseases such as cancer, malaria and AIDS, or by individuals.

Ensure Equal Opportunity for Education in Nation. Equal opportunity is a descriptive term for an approach intended to give equal access to an environment or benefits, such as education, employment, health care, or social welfare, to all, often with emphasis on members of various social groups who have historically suffered from discrimination. Some protected groups include gender, race, or religion. Equal opportunity practices include an organizations' implementation of personnel practices to ensure equality in the employment process. Equal opportunity also applies to equality in housing and public accommodations. In developing countries, equal opportunity is provided by creating jobs accessible to poor people. Economic studies suggest that this is achieved by reducing barriers to entry and allowing entrepreneurial activity.

In the U.S., the federal government and various state and local governments sometimes institute affirmative action in terms of hiring and government contracting with the attempt to overcome alleged historical discrimination against protected groups. There are three different forms of affirmative action, (1) in education; (2) in government contracting and (3) in personnel practices.

Allocate Money to Critical Jobs. In many nations, people are attracted to easy, glamorous, well-paying jobs, such as playing games, singing, or pretending to be other people. These jobs usually pay well above the mean. Many other jobs, that are equally important, such as teaching or police work, pay far less. Other necessary jobs, such as trash collector or store stocker, pay far less than those and have even less status. Perhaps less popular jobs should have higher rates of pay, to ensure that they will get done.

The nation has to try, for itself as an employer and for all employers, to make sure than applicants have equal opportunities to jobs. In many employment contexts, including government entities and federal contractors, not all employers are required to implement an affirmative action program. Employers should compare the incumbency of females and minorities performing specific jobs to the availability of those groups in the recruitment area. Where the employment incumbency in that job group is less than 80% of the availability of that group, the organization should set a better goal. Employers should widely advertise jobs with organizations who might be able to refer qualified minorities or females applicants, although once applicants apply, the employer should choose the most qualified candidate and cannot make employment decisions based on race or sex even where there are affirmative action goals.

Ways of providing equal opportunity are often subject to controversies, since selection or opportunity are difficult to measure accurately. Equal opportunity is said to exist when the outcomes are equal, when people with similar abilities reach similar results after doing a similar amount of work. As long as inequalities can be passed from one generation to another through gifts or inheritance, it is unclear that equality of opportunity for children can be achieved without greater equality of outcome for parents. All applicants should be treated in a nondiscriminatory manner in compliance with established laws, regulations and executive orders. A government should promote a culturally diverse workforce, and a work environment that allows all employees equal opportunity to achieve their full potential, without discrimination.

A government may need to undertake job creation programs in order to assist its people in seeking employment, especially during times of high unemployment or when private sector jobs are lacking or unbalanced—that is, the government may need to create new job categories or emphasize critical jobs, such as teaching or police work. The government can concentrate on making macroeconomic policies to create a supply of jobs or create more efficient ways to match employees with prospective employers.

The transition to an emergency footing will require many new jobs, enough to almost guarantee full employment. The monies to pay for these jobs could easily be reassigned from military spending. In fact, a switch from military spending would free enough money to rebuild national infrastructures, as well as move cities and pay the costs of decarbonization of all industries—and have enough left for free medical care and guaranteed incomes.

Issue National Vouchers at Residents' Birth
A voucher is an economic warrant and guarantee for goods and services. A voucher is a certificate, a document, evidence or proof, of a promise or commitment, which is worth certain rights, materials or a monetary value, and which may only be spent for specific reasons or on specific goods. Examples include, but are not limited to, education, children, and medicine. The nation would either issue these vouchers for all citizens or oversee the issuance by an assigned authority.

Different sets of vouchers could be issued at birth or later in adulthood. Some vouchers might be claimed as wanted or needed.

Basic Income Vouchers for People of a Nation. A guaranteed minimum income is a proposed system of income redistribution that would give each citizen a certain sum of money independent of whether they work or not. In the U.S. this is sometimes known as a 'Basic Income Guarantee (BIG),' 'universal basic income,' or 'guaranteed annual income,' but these systems also often include a method of paying for the income as well.

The system would be a government-administered one that would allot every citizen a sum of money large enough to live on. One amount proposed is 20% of per capita GDP. A basic income voucher could be issued every month, for example, to $7000 a year for U.S. residents. This voucher would cover minimum clothing, housing, and food, although some amount might have to be paid back for carbon taxes or other taxes. Retirement would not be necessary, since payments would continue until death. This program would replace U.S. Social Security. The wealthiest as well as the poorest citizens would receive this. Salaries from employment would be a supplement to this government income.

An often proposed way of paying for this system is through a negative income tax where a government flat tax would be charged to all citizens. The current model of progressive income taxes used throughout the western world could be eliminated, but the system would still be progressive, since those at the lower end of the wage scale would pay less in taxes than they would receive in guaranteed income. For the most wealthy members of society the few thousand dollars of the guaranteed income would only make a small dent in the taxes they would have to pay.

In economics, a negative income tax (sometimes abbreviated NIT) is a method of tax reform that is popular among economists, but has never been fully implemented. It was developed by Juliet Rhys-Williams in the 1940s and later by Milton Friedman in 1962. Negative income taxes can implement or supplement a guaranteed minimum income system.

Proponents of a guaranteed minimum income argue that the system has numerous advantages. First, it would simplify the welfare state. The introduction of a guaranteed minimum income could also see the elimination of traditional welfare, the minimum wage, much of unemployment insurance, government pensions, and benefits for the disabled and ill. This would eliminate large amounts of government bureaucracy. It could also see the elimination of the progressive income tax with no adverse effects on the poor, as explained above.

Next, it would prevent any citizen from falling into abject poverty. With a guaranteed minimum income, *involuntary* starvation and homelessness would be eliminated. And, it might eliminate a few of the major problems of the modern welfare state, such as the welfare trap, which is presumed to discourage people from working.

A minimum income could change the character of work. People would have the opportunity to select other kinds of jobs. It would allow them to do work that is productive but does not normally provide income, such as caring for children or the elderly within one's own family, providing public goods, or working for NGOs. There may be other effects, ranging from increases in employment to depression of wages.

Medical Vouchers for Nation. The American College of Physicians has begun the process of formulating a new voucher system of financing medical care that could ensure universal access to health care in the United States, based on the strengths of existing health care financing mechanisms.

Vouchers could vary in value over time, and may be specialized for use in the separate phases of medical care. They could be individualized for use by a single patient or a single physician. For instance, American Express's new card, Quattro, will be used by 500 of its workers and those of John Hancock Financial Services in the Boston area.

A double voucher system for patients and physicians could provide mechanisms to assure that the care decisions reached between patient and the physician are not subject to prior review or micro-management by the insurance carrier, which has been a cause of the increasing frustration and anger felt by physicians and patients over denials of care. The voucher solution will go a long way to restoring the beneficence needed for everyone in the medical interaction.

There are a number of ways a voucher could work; it could be good for 24 months of treatment, or for a lifetime. It could also be traded. Birth and death services might be free.

Productivity & Resources Vouchers for Nation. Here is one possible way the numbers could be determined: After calculating an optimum population for the planet (based on totals of ecosystem productivity and resources), that number is divided into the estimated quantities for each resource on the planet. Based proportionately on territory, each nation is given a number of vouchers for the optimum use of resources and processes. Nations may then choose to set their populations accordingly, or to make decisions based on other values. People would have access to a share of the total natural plant productivity in an area. This would replace an idea of straight calculation of 2.5 hectares per person. And, that would be after other land uses were subtracted.

Carbon & Other Elements Vouchers for Nation. Every nation would be issued a number of vouchers equal to the carbon-bearing products of its population and their standard of living, based on a world standard. These would determine how much was used in each national system, regardless of its population or the intensity of its exploitation. Other elements, such as copper and sulfur, would be treated the same way.

Vouchers for Water, Air & Other Compounds. The requirements of living, such as breathing or drinking, should not require vouchers, except under extreme circumstances of overpopulation or overconsumption. Trying to issue vouchers for these things, on a personal level, might be too expensive and too invasive, although assigning the vouchers might protect basic needs.

Education Vouchers for Nation. National educational vouchers could be issued for skills that are in short-supply or demand. Education would not be required per se, but it would be required to have a driving or voting license. The nation would reserve the option of requiring the first six to twelve years of education, but any individual could claim the education later. A nation could issue other vouchers for post-secondary education, especially to emphasize areas where skills were needed.

Reproduction Vouchers for Nation. A nation might have to issue child or reproduction vouchers, depending on the self-reliance or choices of its communities. If communities were balanced at the national level, then the vouchers would be controlled at the national level. This would allow some communities to have more or fewer children, based on their suite of values. China has something like this in place now, that is, there is a limit of one or two children per family, although children can be placed with childless relatives. Furthermore, the most of the Chinese people recognize that places are overcrowded and think this is fair.

Other Vouchers for Nation. All the previous vouchers may be traded. This may be necessary in the case of partial vouchers. Should there be vouchers for luxury? In a random lotto system for fairness? Or should it be a reward for generosity?

Special Accounts for a Nation
In common usage, saving generally means putting money aside, for example, by putting money in the bank or investing in a pension plan. In a broader sense, saving is used to refer to cost-cutting costs or to rescue. In terms of personal finance, saving means to preserve money for future use, by putting it in a safe place, such as a mattress or banking institution. Saving refers to an activity occurring over time, a flow variable, whereas savings refers to something that exists at any one time, a stock variable. The flow refers to an increase in assets, whereas the stock refers to a designated part of the entire assets.

 For a nation, special savings accounts would be set up to lessen the extent of catastrophes or emergencies. One of those would be to keep stores of food to feed each community for as long as possible, or at least for seven years. There would also be accounts for helping other nations and for paying dues to the Global Union organization.

Start National Savings. A nation needs to start savings accounts for lean years. Each nation should establish at least one account for food and one for nonrenewables. Perhaps each account should hold a seven-year supply. Genesis in the Christian Bible provides the model: "And the food shall be for a store to the land against the seven years of famine, which shall be in the land of Egypt; that the land perish not through the famine." Some nations may prefer to set aside money, although food and supplies may be more certain than the value of money during a world-wide famine.

Start National Account for International Charity. A nation should also keep accounts to help the wildlife and people of other nations. This would not only provide safety nets for other nations beset by temporary problems, it would promote trust between nations.

Pay National Share of Global Costs. Every nation would be expected to pay a share to support the Global Union of Commonwealths (GU). Since the GU would be self-supporting, with its own forms of income, this would be a form of membership dues and national dues would not be required.

Discussion of National Payouts
A nation could afford to make these kinds of payouts because of the restructuring of income. In fact, just suspending many subsidies and handouts would free billions of dollars for many nations.

Remove Legacy & Current National Subsidies. Many governments subsidize activities that threaten their own sustainability. Lester Brown points out that fishing fleets are subsidized at approximately $54 billion per year (2000), despite the fact that fishing capacity exceeds the capacity of the ocean to yield sustainable fish. Norman Myers and Jennifer Kent placed U.S. agricultural subsidies at $390-520 billion (USD); $110 billion to fossil fuels; and $220 billion for water. All subsidies are over $2 trillion.[13] By comparison if all U.S. citizens received $7,000 per year in negative taxes that would cost under $2 trillion.

Another example is the costs of modern industrial farming in England: $2 billion (USD) for removing pesticides from drinking water, damage from soil erosion, medical costs of poisoning and mad cow disease (which is 90% of what farmers earn); $4 billion for subsidies to farmers (180%); $1 billion for health care costs for poor choices (45%); and $3 billion for loss of productive land.

One action that policy makers could take to meet tax reform or deficit-reduction goals is to eliminate a number of existing tax exemptions, deductions and loopholes that are both fiscally and environmentally damaging. Most notable among these are provisions for mature industries such as oil, mining, timber and automobiles. Not only do these government handouts to corporations work against common-sense notions of free markets, innovation and fiscal responsibility, but they also encourage mounting environmental and health costs that society must bear. The U.S. public is paying billions of dollars a year to help corporations make profits on things that harm the public and the environment.

Nations should remove most subsidies, especially for fossil fuel, farming, fishing and roads. Temporary subsidies might be used to encourage solar power or wind power, or for hydrogen use to replace fossil fuels in transportation.

Disallow National Tax Cuts. The bulk of tax cuts, under the graduated system in the U.S., for instance, disproportionately benefit the wealthy and are meaninglessly small for everyone else. With the increases in use and loss taxes, and the elimination of income taxes, there are no reasons for tax cuts (although the small minority taxed for heroic incomes may disagree).

Organize & Manage Complete Economy for Nature and People

An economic system is a mechanism (or social institution) which deals with the production, distribution and consumption of goods and services in a particular nation. The economic system is composed of people, institutions and their relationships. It addresses the problems of economics, such as the allocation and scarcity of resources. It has to address the entire context, also. The equal apportionment of 'resources' to all cooperating participants in the global commons (which could be identified with the new word 'Koinomics') is supported by the theory and practice of recognizing and honoring the legacy of the entire planet where every being is a tenant.

Encourage Efficiency. A nation can encourage efficiency. Allocative efficiency is the market condition whereby resources are allocated in a way that optimizes the net benefit attained through their use. Allocative efficiency is also defined as the production of the quantity that is most beneficial to society.

The term eco-efficiency, identified by the World Business Council for Sustainable Development (WBCSD) in its 1992 publication 'Changing

Course,' is based on the concept of creating more goods and services while using fewer resources and creating less waste and pollution. The 1992 Earth Summit endorsed eco-efficiency as a means for companies to implement Agenda 21 in the private sector. According to the WBCSD definition, eco-efficiency is achieved through the delivery of competitively priced goods and services that satisfy human needs and bring quality of life while progressively reducing environmental impacts of goods and resource intensity throughout the entire life-cycle to a level in line with the estimated carrying capacity of the planet. This concept describes a vision for the production of economically valuable goods and services while reducing the ecological impacts of production.

According to the WBCSD, critical aspects of eco-efficiency are: A reduction in the material intensity of goods or services; a reduction in the energy intensity of goods or services; reduced dispersion of toxic materials; improved recyclability; maximum use of renewable resources; greater durability of products; and, increased service intensity of goods and services. The reduction in ecological impacts translates into an increase in resource productivity, which in turn can create competitive advantage. In 2002, Michael Braungart and William McDonough expanded the ideas of eco-efficiency and its practical applications.

Certify National Standards. Consumer pressures have often influenced the extent of exploitation in the national landscape. Consumers have indicated, for instance, that they want forests for wood products, but also for habitat for other species, aesthetics, recreation, and other values. Certification itself is a demand. As consumer demands change, however, the uses of landscapes may change. Certification may bring about other far-reaching changes.

Certification is one way for consumers to discriminate against products that do not enhance other values. It can be applied to many processes. Some may see certification as a way to protect ecosystems and their nonmarket benefits, such as cleansing water and air. Others may see it as a way to sell items for more money, or as a way to avoid the social conflicts over limited resources. Certification is a way for all stakeholders, including consumers, to define and refine demand. But, it cannot be isolated from the large-scale economic and political trends that shape human culture. Problems are part of a matrix of industrial social practices and policies, but there are now pressing economic and ecological reasons to revamp them.

Certification will require more knowledge about ecosystems, as well as improved inventory and monitoring levels and techniques. Certification implies that a complete inventory must be taken of the system to provide a baseline for further assessment. It also implies that monitoring must be continuous. Certification requires much better scientific research to determine minimum, optimum or maximum numbers of features for ecosystem health.

Certification will address concerns of human equity and social justice. Land tenure is one barrier to meaningful certification in many parts of the world. Poverty is another such barrier; as long as the compensation of workers depends on the style of the ruling regime, there may not be enough stability for certification. Certification can contribute to a needful awareness among political representatives, who have been hypnotized by the flow of money. Certification has the potential to increased desperately needed revenue for regeneration, management, and research. It has the potential to increase revenue to local communities, rather than contributing to the flood of cheap raw materials exported elsewhere. In a very real way, it can encourage community cooperation to create millions of jobs and generate billions in expenditures.

Living generations are responsible for limiting their actions within a reasonable framework of cost and irreversible change. The standard requires conversation and some consensus about the limits, which are never exact. Ecological value has to be balanced with optimal resource allocations.

Certification can identify issues that must be examined to bring about self-sufficient societies exploiting renewable resources. The concept of sustainability is being incorporated into every level of planning and management. The concept can be used to develop practices that decrease depletion and waste, and reduce the threats to future generations.

Certification for Forestry. Certification is the provenance of wood made visible and known. It is part of a complicated network and has far-reaching implications. Certification is driven by consumers, but consumers are reacting to forest degradation and destruction, which are partly caused by bad forest practices and by consumer demand for wood products. Consumer pressures have often influenced the extent of exploitation in a landscape. Consumers have indicated that they want forests for wood products, but also for habitat for other species, aesthetics, recreation, and other values.

There are several independent groups offering forest certification, ranging from the forest industry to conservation and national groups. These groups need to be certified at a national level to ensure minimum standards.

The language of certification must address economic and management concerns, but it can never forget that the forest is a living, breathing being, and not a resource or a machine. Certification may bring about other far-reaching changes.

Certification for Energy Generation. All energy generation needs to be certified, from oil wells to oil rigs. Fracking, for instance, was rushed into production before adequate environmental regulation and monitoring; and with zero efforts at long-term impacts. Fracking is exempt from 6 key pieces of legislation related to hazardous waste and pollution. Pro-fracks in the US paid $239 million for candidates and 726 million for lobbying. The EIA overestimated natural gas production. Each fracking well uses 9-29 million liters of freshwater over its lifetime. In the US much of wastewater pumped back into wells or waterways. Many countries, such as Italy and Bulgaria, have already banned fracking. UNEP states that even done properly, fracking will have unavoidable environmental impacts. The US government is avoiding its responsibility for protecting citizen health from fracking. Low-energy prices are short term and do not balance the long-term damage. Costs for health and clean-up are going to be high. Fracking is also a threat to climate stability.

A shift to efficiency and alternative sources is better strategy, but fracking is a roadblock to expenditures on alternates. The UK Tar Sands Network shows how everyone on planet could have a good quality of life without fossil fuels at all, with fair sharing and renewable sources.

Calculate an Optimum Population for the Nation
Ecological design uses five general steps to calculate a population, for Ireland as an example. The first is to identify the place within its natural boundaries. Ireland, for example, is a uniquely identifiable island ecosystem, with recognizable boundaries and a unique history and character. It is part of a larger regional context, with neighboring landmasses, as well as oceanic and wind patterns. All systems on the island should be inventoried and monitored. Second, an optimum amount of wilderness to preserve the natural cycles indefinitely should be calculated. If the current area is less than the calculations, as it most likely is in Ireland, the difference should be restored and set aside as a reserve. Third, the remaining area would be zoned for appropriate use, including conservation, preservation, and artificial areas, with historical, cultural, and functional importance. The infrastructure, including agriculture, energy generation, and urban areas, could be reworked to be more appropriate and efficient. Fourth, we need to identify the resources available for human use, including raw materials and the biological productivity of the areas (this productivity is necessary to calculate an optimum population). Then, we need to plan the long-term use of resources, identifying appropriate and best uses. And, fifth, we need to apply cultural modes—in style, values, and technology—to suggest limits on technology and population. This would preserve cultural values. Encouraging popular participation in an open design would allow people to invest in the design.

As part of the formulation of a design, it is crucial to examine the natural and cultural histories of Ireland. We need to understand interactions in the ecosystem, as it existed before humanity, as it was lightly settled, and as it is now, dominated by humanity and coevolving with culture.

Just as important as any calculation of the population in an area is the rate of resource use by that population. Grasslands and forests have been altered thoroughly or devastated by human occupation, by overcultivation and overgrazing. Technology can be used to expand or contract resources. Technologies have the capability to minimize the use of resources, but they also have negative effects.

Is 8 million people in Ireland too many? Is 5 million? How many more people could live in Ireland sustainably? 4 million? 1 million? Is this number above or below some maximum carrying capacity? There is a maximum carrying capacity for this region, as well as an optimum. The carrying capacity is the largest population sustainable on a long-term basis of renewable and nonrenewable resources. For humans, this capacity must include domesticates, as human equivalents, since many domesticates compete for protein consumption. Domestic animals can extend the carrying capacity somewhat, since many of them consume agricultural wastes or use lands marginal for agriculture, but they are not as efficient as wild populations, which could be harvested more lightly.

Technology can expand the carrying capacity to some extent, with higher yield crops and resource substitution, but also it reduces the capacity with unforeseen effects, from the use of pesticides, for example. Furthermore, the capacity decreases as the per capita use of energy and resources increases. Carrying capacity calculations often just consider food energy, but all needs — clothing, shelter, transportation, information generation, and aesthetic satisfaction — must be included.

Eugene Odum suggested using land area as a measure of human carrying capacity. Using Odum's technique for Ireland, assuming that wilderness area has been considered in the calculation of natural areas, and converting for the differences in productivity of ecosystems (about the same as Georgia), the population calculation comes to 3.4 million people. This figure, however, does not include trade land, that productivity needed to trade for necessities, such as rare metals; this might reduce the number to 3 million people. It is possible to calculate an optimum population using Net Community Productivity. For a combination of temperate grasslands and forests the number is much lower. The maximum population calculation for NCP results in 2.4 million, *with an optimum of 1.9 million.* The population in the 1850s may have reached 8 million; after the potato famine and subsequent deaths and immigrations, the population dropped to 3 million. The Irish population is approximately 6.2 million in 2010, including both the Republic of Ireland and Northern Ireland. This is a significant increase and it may not be sustainable at its growth rate of 8 percent.

Represent All People in a Nation
A nation is obligated to represent all people with its borders, even those who do not have formal citizenship. A nation is also obligated to guarantee resources first for its authentic citizens.

Obligations to Protect Rights & Privileges in a Nation. The rights that have been established need to be protected, not only from impingement by other nations or a global association, but from changes within the nation itself.

Obligation to Meet Basic Personal Needs in a Nation. A nation has an obligation to meet the needs of its people by planning for enough food, land, resources, energy, and opportunities through growing or trading, and by planning the development of the structure or the national population.

Obligation to Integrate Everyone into a Nation. The nation has to be able to integrate the people into a national population, through acceptance, education, and shared values and actions. It has to make sure than no conflict between groups or individuals becomes out of control or involves large numbers of others.

Work with Other Nations
A nation has to acknowledge the existence and rights of other nations. It is no longer possible to ignore another nation, although every nation has the right to limit the number of connections and exchanges with other nations. So much, in terms of global patterns or trades, is shared among nations that a nation is obligated to work with another nation if there are common problems that need to be resolved.

Assess National Performance through Surveys
A nation has to receive feedback on its performance, from distributing resources to protecting rights. This feedback can come from other nations, its citizens or its representatives. A nation can monitor its own activities, to ensure that standards are followed for the duration of the activities. Monitoring also allows the standards for the activities to be revised to increase efficiency. The results of monitoring need to be evaluated and fed back into the process of planning and acting. The skills and knowledge of those who do the evaluation determines the success or failure of the principles and standards developed in the plan for monitoring and assessment. Interactions need to be evaluated in terms of impacts, needs, goals, and limits. The whole system needs to be evaluated.

Preparing Leadership
Isolated problems stir only mild attempts at reform. But, few problems are isolated. Modern government is the assumption of responsibility for problems without adequate knowledge. Modern citizenship is the abandonment of responsibility, on the assumption that others have adequate knowledge to manage problems. But, our leaders refuse to learn or understand the simplest facts of science or technology. Leaders have to acquire an ecological consciousness.

Many adjustments may be necessary at a national level, such as civil service tests to determine candidacy, as well as board review for basic minimum moral and social qualifications. Making these changes all at once might seem unrealistically disruptive, but would it be? Would it be more disruptive than an earthquake or a war? Would it fracture the cultures or communities? Would people's lives change that much? Would the change be worse than a lost job, a business failure or death in the family?

These things should always be done if people were convinced by their respected leaders that this was the best way to proceed. Given the fact that the characteristics that make people lust for leadership often detract from their abilities to lead, perhaps we need a form of draft to get leaders, or perhaps we need to consider some who have been groomed for decades or from birth, like the Dalai Lama.

There must be other ways to keep elected or appointed leaders balanced. Perhaps each should be assigned a jester, which would be an appointed office. We need to institute that office again, but maybe without the possible death penalty. In the comedy of politics, the jesters would present common sense and truth. A new Office of Jester is crucial to the continuity of modern government. Politicians have learned that nothing is too dumb or extreme, regardless of how ridiculous or socially destructive it is. As long as a few of their constituents love their idiocy, cute greed and smiling shrewdness, they continue it. Since politicians likely do not see comedy shows or read the comics, it might help if they had a personal Fool to remind them of their intellectual and moral errors, not only to protect us from losing our health and paltry wealth, but to protect them from a weak conscience or moral meltdown.

Jesters had to be amusing, but were expected to be honest in observing and commenting on the behavior of their betters. They had the power to mock and revile the most prominent. It is urgent that we create a formal position of Jester to attend every high political position, especially in courts, and ground it. We should not just fool around. It might cost something, but it would employ many unemployed comedians. They could lightly amuse the leader and then deliver the really bad news: Shrinking economies, social upheaval over massive inequities, dangerous armaments not for warfare, and runaway environmental destruction and looming catastrophes. Basically the message is: Stop pretending that all is right and that business as usual can save the day—It's an emergency!

Until we can do things naturally and spontaneously, we must try to act our way into right thinking, while thinking our way into right action. It is a form of intelligence, to know the principles and trends of natural affairs so that the least energy is used in dealing with them. This is also the unconscious intelligence of whole organism and the whole community.

Figure 4. *Map of Proposed Restored Forests in Ireland*

Contributing to the International Level

A Eutopian framework could be implemented immediately. Most global studies, such as those from the Club of Rome and *The Ecologist*, state or imply that change cannot be fast, that people cannot adjust, that social disruption would result, and that chaos would finish what ignorance and technology could not. These studies propose slow, long-range plans, while warning at the same time that the earth is facing imminent, drastic change. If their plans are implemented too slowly, and if the population or pollution doubles again, surpassing some unrecognized critical level, there would be worse disruption. The Eutopian proposal is for immediate action, scaled on a week or a month.

Sudden change is already a hallmark of industrial progress. Industrial cultures have replaced older patterns with great suddenness. Eutopias cannot seem more sudden than the loss of a home or place. Industrial cultures have reduced people's control over the means of production and power. Eutopias does not offer less control. Whole communities have been destroyed by industrial scale. Our social structures are already changing rapidly and impractically.

Let us take immediate action to make the changes conscious and more practical. Eutopias offers movement towards common, achievable goals. Eutopias would be a framework for cultures, where different human experiments are tried. Its variability would insure that we could reject any of the local visions that fail. People may object to giving up too much or not gaining enough. Eutopias may be called anti-human, anti-progress, anti-scientific, anti-technological, or anti-educational, but it is merely a new framework for conducting traditional human activities.

Natural environments and human societies are wobbling. Many contradictory impulses are leading to unbalance; some countries want to consolidate into economic powers and others want to secede into independent units. Human civilization will tear itself apart if we let it. We can slow it down and direct it.

The need to maintain our comfortable status for as long as possible, fatalism that nothing can be done or it is too late, prejudice, ignorance—all are keeping us from moving. There are other reasons not to move: Failure of knowledge, failure of communication, failure of imagination, and failure of nerve. Much human suffering is caused by self-deception, which leads to isolation and then anger, reaction, and more suffering. Real change is difficult in this state, but change is more difficult for people who are starving or oppressed.

For most people in agrarian countries, even freedom from hunger and sickness is utopian. For most people in industrial countries, the choice of a fulfilling profession is utopian. Grinding poverty, economic dislocation, homelessness, are more painful than a transformation to Eutopias. Already most cultures have been transformed by cash crops, mining, tourists, highways, high-rise housing, and condominiums. Physical disruption has been more extensive than the transition to Eutopias could cause.

A long view seems meaningless when so much suffering already exists. An immediate, realistic, coordinated program of action is needed, capable of being implemented by communities and global agencies. We must face our responsibilities directly, declaring that there is no place in a eutopian society for monopolistic and multinational corporations, for the maniacal religion of merchandise, for genocidal military establishments, for urban explosion, for state socialism, for overbearing bureaucracies, or for technocratic politics, we must act to end them. The declaration must be political, through cooperative networks or leaderless consensus, by persuasion and example. The problem of human existence on the planet must be approached without deference to artificial boundaries of states, races, or castes. Poverty, pollution, repression, are concerns of every human community. We must stand and state that nature has limits, that we cannot have all we want.

The application must be immediate. The crisis of exponential growth and destruction cannot be solved just after some final limit is approached or passed. The crisis of ignorance cannot be solved by hurrying ahead and creating more problems. Paradoxically, the best thing to do is stop—stop growing, stop producing, stop running; suspend the race and contemplate a direction. We have been asking how the earth will survive its human populations and how they can be lowered. Let us just freeze growth and see what happens. Let us just freeze the populations—a year or decade of no births. We know that whole countries have lost a generation and continued. We know they have rebuilt again from ruins. We could build from recycled materials alone, so there is nothing to fear from stopping. Immediate social reforms, the reallocation of resources, and the preservation of wilderness are necessary, because of the nature of the problem; we cannot predict global climatic or ecosystemic catastrophes. Substantive change and research cannot be delayed until academic or political controversies are resolved.

The transformation must be complete; it cannot be done partially. Global political and economic institutions must all be changed. The Global Union (GU), more empowered than the United Nations, must have authority for the preservation of nature and human cultures. Holistic change will permit the reorientation and balance of local institutions. For example, air pollution is not independent of industrial processes, transportation, and employment patterns. Communities must be of a size that their members can feel responsible for them. These changes are demanded by new requirements, ecological balance primarily among them. New institutions must be compatible with these new values.

The approach must be pragmatic and flexible. By its nature, the Eutopian frame could reduce some of the stresses of transition, the uncertainty, ambivalence, or reversion. The readjustment to the realities of our new intricate involvement in the whole order of nature and her ecological balance will cause social swains. Some capital of energy and materials may be wasted. Population will be matched to solar budgets or net ecosystem productivities. Production will be redirected to communal needs in transportation, housing, food, and recreation.

There will be problems regarding the breakup into more natural cultural divisions. Some will want to decide boundaries by ecosystem; others through culture, watershed, or political power. The Global Union will have to decide when two groups claim the same place or when cultures combine through unions and conspiracies.

There will continue to be other problems. Cutting trees in Nepal causes floods in Bangladesh, and floods cause deaths because overcrowding has forced the poorest people to live on flood plains. The poor in the highlands everywhere effect those in the lowlands, often adversely. The quest for ecological balance means that some ecosystems must be maintained by systems managers, who often overmanage. The larger the human impact, the more control is necessary, until balance is restored. Eutopias seeks to improve people's circumstances by enlisting them to save their environment and their way of life.

People cannot be given material equality instantly. But things can be leveled within a culture; cultures with excess may be taxed by the Global Union. Providing work for everyone is one way to narrow income differences. The GU, nations, communities, and families must provide it. Worthwhile work requires imagination. The large work force employed by military contracts in industrial countries will be dislocated at first, but that employment is supported by taxes, which could be reallocated for construction and deconstruction—so many highways, manufacturing plants and abandoned buildings.

Crime and civic unrest will not disappear. The Global Union and nations could reduce many kinds of global and victimless crimes with new policies. Because most cultures have strong policies regarding drugs, abortion, and prostitution, among other things, the GU would not impose rules on every crime. Dangerous weapons, from automatic guns to tanks, and dangerous products, including nuclear reactors and biocides, would be strictly regulated.

People will still make mistakes and bad choices in a Eutopian framework. But, if a form of government is bad or ineffective, then it can be altered more easily in a smaller, more flexible framework. In the Eutopian framework, people can learn from mistakes or unintended side-effects—as when doing good that causes evil. The scale is small, so the catastrophe is small. There will always be some injustice, inadequacy, and unpredictability. Large political and economic institutions have only made it worse. If Eutopias turns out not to be the proper framework to solve these problems, it might lead to a better way.

Implementing Immediate Steps for an International Body
The GU could be based on the UN, but with immediate responsibilities and powers, for protection and preservation, as well as some temporary powers, such as taxation. Too many things have happened in the past 100-500 years. The solutions need to be started now. Immediate steps are necessary to address catastrophic changes.

Creating an International Body
The original charter of the United Nations (UN) restricted its activities to peace-keeping and human rights. While these are still important, other things need to be addressed, especially as they directly relate to concord and harmony (peace and health). Many things need to be done. New organizational bodies need to be created. These bodies would operate inside the new Global Union (GU) and be harmonized with existing bodies. Other bodies, such as the World Bank or International Monetary Fund, that operate parallel with the GU, must be reintegrated as part of the GU Financing. They must be governed by the GU executive branch, rather than by their own independent governing bodies, which are unduly influenced by wealthy countries. The GU needs to control it agencies and departments, as well as its financing.

The GU must have the power to define itself, to rewrite its charter so that it encompasses more than security and health, or expand those things to include all of human activities. Several studies of the UN, such as the Jackson and Pierson reports, have already focused on organizational efficiency — and efficiency has been the dominant factor in all bureaucracies for the past one hundred years — but the organization needs to be relevant to address the challenges that modern global trade has proposed to traditional cultures. The North American Free Trade Agreement (NAFTA), the TransPacific Partnership, the TransAtlantic Trade and Investment Partnership, and others all work to undo many national environmental and labor laws. The GU has to limit the power of these agreements and uphold the right of nations to protect their environments and workers.

The GU can accept new nations as members. Membership in the GU needs to be opened to those states now represented by UNPO, that is, cultures, without recognition of their land, as well as other cultures that may not be landed at all. Recognition may also be granted to the regional association of cultures or nations.

Independent cultural areas within nations shall have the status of independent nations within the GU. Any culture would be given legal recognition, protection, and full autonomy over their boundaries by application to the GU, which would determine priority of claims (by archaic peoples, agriculturalists, pastoralists, or industrialists). No action would be taken to disband existing nations. Nations could still remain allied with old, larger nations as independent or dependent regions, although they would have only one vote at the highest federation. The nations would determine the use of allocated resources. Local economics and technology would provide for populations. Traditional religions and customs are maintained or permitted to develop.

The GU may consider whether it should extend recognition to voluntary economic associations, that by reason of size or power, such as some international corporations or regional industry associations, have great influence of capital and resources. Once membership is nonculturally or nonnationally based, however, it must be open to NonGovernmental Organizations, that may be dedicated to special interests or to preservations.

And, then, perhaps the interests may be religious or spiritual. However, for voting purposes, nations or organizations may have to give up privileges as one or the other. On the other hand, the GU might require corporations and organizations to be based at the community or national level and participate in them materially.

The framework would allow a loose unity. For example, consider this lesson from fifteenth-century China: Too much unity stifles a creative advance; or this lesson from Europe: Too much disunity wrecks a creative advance. Europe had enough disunity to prevent unification, but not enough barriers to prevent the spread of technology and competition for technology. The role of the GU would be to provide just enough unity for nations to advance and develop. A Eutopian framework is a form of 'balkanization' with limited barriers and limited unity. The barriers serve to protect the culture, its resources, land, its people and standard of living.

Having a Different Structure. The GU would create a different structure than the UN. The GU could be modeled as a federal government, although it would be composed of independent nations. This would be necessary due to the responsibility of the UN for global issues and to coordinate nations. The GU would also guarantee fundamental environmental nonhuman rights, as well as human rights based on common things.

The GU would rework its constitution as a formal process of expressions and expectations. Traditionally a constitution is an agreement to divide into rulers and ruled. For the GU it would claim responsibility for all global affairs, including the environment and large weaponry.

The executive branch, the Secretariat, would be elected from the General Assembly. Seven to twelve members would be elected. They would elect a coordinator with the title of Secretary or Coordinator. It would have authority over all departments and agencies. It would mange the organization as an organic whole. The General Assembly, as a legislative branch, would make global and international laws. It would strengthen the Declaration of Human Rights (1945). It may make the planet into an incorporated entity. The judicial branch would be the International Court

Other branches may be separate and equal. Nations would nominate others for these branches. A Commission on the Earth would have the responsibility to consider all events in the environment from earthquakes to biodiversity. Other important Commissions would be the Commission on Beliefs and Religions and the Commission of the Heritage of Cultures.

The Office of Ombudsman would be expanded and strengthened. Although not traditionally a fourth branch of government, as first proposed in Sweden, its function would be to improve the effectiveness and fairness of the global government while protecting basic human and ambihuman rights. One strength would be its power to initiate legal proceedings in behalf of any human and ambihuman constituents or nations. It would be an independent branch with separate responsibilities: To investigate complaints against the UN or its officials for wrongdoing; to investigate complaints against nations; and, to investigate the operation of the UN.

Start Catastrophic Measures by the International Body

Because the challenges are immediate and because the consequences of not meeting them could be catastrophic, these things have to be initiated immediately. Of course, they are part of a process that may never be complete, like a one-time fix, but they have to be started now.

Transfer Powers to the International Body
The major military powers would grant their powers to the Global Union and relinquish their efforts towards global leadership; they also resign from the security council, cease propaganda activities, renounce foreign policy objectives, call back all soldiers from foreign countries, and stop giving away produce, factories, or weapons. They put their technical and educational surpluses at the disposal of the GU. If the USA, China or Russia is to be a world leader, let her lead in tolerance or in trust. Let her be the first to give allegiance to a world organizing body, the GU, the first to divest themselves of nuclear weapons. If they fear for safety, they need only remember the success of nonviolence in India or of guerrilla actions in Southeast Asia, Central America, or the Middle East.

Disarm Nations by International Body
In a separate essay entitled "Toward a World Social Contract," Kenneth Boulding examined possible global mechanisms to abate conflict within "Spaceship Earth," including "universal policed disarmament down to internal police levels" and "the organizational union of the armed forces of the world under a limited world government" — both of which are key elements of unfolding official U.S. government disarmament policies envisioning a world "effectively controlled" by the United Nations.

The 1945 UN Charter envisaged a system of regulation that would ensure "the least diversion for armaments of the world's human and economic resources." The advent of nuclear weapons came only weeks after the signing of the Charter and provided immediate impetus to concepts of arms limitation and disarmament. In fact, the first resolution of the first meeting of the General Assembly was entitled "The Establishment of a Commission to Deal with the Problems Raised by the Discovery of Atomic Energy" and called upon the commission to make specific proposals for "the elimination from national armaments of atomic weapons and of all other major weapons adaptable to mass destruction."

The UN has established several forums to address multilateral disarmament issues. The principal ones are the First Committee of the General Assembly and the UN Disarmament Commission. Items on the agenda include consideration of the possible merits of a nuclear test ban, outer-space arms control, efforts to ban chemical weapons, nuclear and conventional disarmament, nuclear-weapon-free zones, reduction of military budgets, and measures to strengthen international security.

The Conference on Disarmament is a forum established by the international community for the negotiation of multilateral arms control and disarmament agreements. It has 66 members representing all areas of the world, including the five major nuclear-weapon states, the People's Republic of China, France, Russia, UK and U.S. While the conference is not formally a UN organization, it is linked to the UN through a personal representative of the Secretary-General; this representative serves as the secretary general of the conference. Resolutions adopted by the General Assembly often request the conference to consider specific disarmament matters. In turn, the conference annually reports its activities to the Assembly. The conference indicates that disarmament ideas are considered.

The U.S. must keep its own promise to give up nuclear weapons, which it made in 1970 when it signed the Non-Proliferation Treaty. The U.S. must terminate its $8 billion annual program to develop new weapons and agree to Russian President Putin's offer to reduce the mutual nuclear arsenals of about 10,000 weapons down to 1,000, or down the ante to zero.

The U.S. and Russia need find parity with the arsenals of the other nuclear weapons states, China, UK, France, and Israel, with stockpiles in the hundreds, and India, Pakistan, and North Korea each with less than one hundred bombs, should find reduction easier. China has offered to negotiate a treaty to eliminate all nuclear weapons and call every nuclear weapons state to the table. There already exists a plan for a model treaty prepared by scientists, lawyers, and policy makers, which was submitted to the UN as a discussion document. It lays out all the steps for dismantlement, verification, guarding, and monitoring the disassembled arsenals to insure that we will all be secure from nuclear break-out.

Russia and China have repeatedly offered a proposal in the UN General Assembly to ban weapons in space. That is the only way that those two nations will agree to join the U.S. in abolishing nuclear weapons. They do not want U.S. control and domination of space, which indeed is the aggressive mission statement of the U.S. Space Command, to gain military superiority over the whole planet. We must also replace the Non-Proliferation Treaty's guarantee of an 'inalienable right' to so-called 'peaceful' nuclear technology. This is the provision on which Iran is now lawfully relying. It could be nullified by establishing an International Sustainable Energy Agency and phasing out nuclear power. Every nuclear power plant is a potential bomb factory, so it would be impossible to eliminate nuclear weapons without eliminating nuclear power.

An international body could disarm all nations; it could fund its work through the redistribution of the $200 billion in tax breaks and subsidies given to the nuclear and fossil fuel industries world-wide. The treaty negotiations and actual dismantlement of the nuclear arsenals could be done within weeks, although years or decades will probably be requested. If nuclear weapons were illegal, Korea, Iran and other nations might be willing to give up new programs, without recourse to attack or bribery.

All other nuclear weapons nations have indicated that they would be willing to give up their nuclear weapons, numbering in the 100s or 10s, if the larger nations do. The U.S. has been the biggest block to the efforts to stop proliferation. It is naive to believe that anything less than the elimination of nuclear weapons will reduce the possibilities of nuclear war and the unimaginable catastrophes that would follow a nuclear winter.

Elimination of nuclear weapons will not stop wars, although it is less likely that wars could be as destructive or final. Other weapons of 'mass destruction' also need to be reduced and controlled, especially long-range guns or automatic missiles. These weapons are designed to kill as many people as possible as quickly and cheaply as possible, regardless of their status as combatants or civilians. The International GU could work to bring every arsenal to parity.

Take or Destroy Nuclear or Large-scale Weapons
Complete disarmament could be accomplished within a week. Earl Osborn proposes this concept of sudden disarmament in response to the tedious phase-out envisioned by most plans. An agreement would not involve much negotiation. Taking this first step would add to the prestige of the country bold enough to do it. The GU could post a police force to disable all military ordinance. A thousand planes each carrying one hundred trained inspectors could be distributed at all major centers in the nuclear countries within 24 hours (1 day).

Allow Personal Weapons in International Body
For Nations, the GU might make available cruise missiles, for example, at some agreed-upon level, to any nation, as a maximum defensive tool. For individuals, it might allow personal weapons from hands and knives to arrows and single-shot guns, depending on national choices and limits.

Start Year of Consideration through International Body
The Global Union promotes a year of consideration. Starting with population growth, all economic growth, or expansion of claims, would be suspended for a year. Age grouping would be discontinuous for a while, if further years of consideration were necessary. The use of certain poisons or chemicals, especially greenhouse gases and other chemical additives that destroy ozone, would be discontinued.

Start Equity Measures
The global environment needs to be equalized also. If every nation or person has certain rights to land and resources, then those have to be respected and adjusted. Some GU taxes would be designed to repair the inequity from hundreds of years of unfair trading or accumulation.

Implement Global Ecological Goals
Global goals apply to the planet as a whole, for Gaia as a metaphor for the living planet. These goals are not simply the sum of local and regional goals. The GU would:
- Reimplement international initiatives to slow deforestation – the

UN notes that previous initiatives *accelerated* deforestation, as in Cameroon, where log production is planned to double in the forest, home to 50,000 Pygmies with a unique and valuable life-style
- Plant and maintain forests sufficient to guarantee indefinite support of known and unknown global biogeochemical cycles
- Protect fragile ecosystems with global importance
- Reduce threats to ecosystems from acid rain and other nonpoint-source pollutions to less than 5 percent of present values
- Plant 9 million ha of trees each year to meet current demands; for soil and water conservation, plant another 6 million hectares (at an estimated cost of $6 billion dollars); and plant 110 million hectares just to catch up with cutting
- For the planet, reforest 1.4 *billion* hectares to restore the 30-40% forest cover removed in the past 3000 years.

The GU would monitor ecological goals for nations and communities.

Implement Global Cultural Goals

Global cultural goals apply to every human culture.
- Protect the health of human communities; set standards for health and provide information and assistance to nations.
- Provide educational assistance to nations. Encourage cultural education and history. Encourage teaching of local languages. Allow renewal of critical customs. Educate all people to feel their connections to their place, because, until they feel them, they will not act ethically or ecologically. Educate people to realize that long-term sustainability requires healthy places, and that protecting places protects jobs and values.
- Stress education for women, especially in nations who are still denying it to them for ideological or selfish reasons.
- Coordinate and support economic activities. Broaden local economies from resource extraction to invention and specialization. Promote the full use of ecosystem products; support small-scale businesses that produce new, high-quality products. Increase manufacturing efficiencies.
- Guarantee sovereignty for nations. Offer a platform for communication with other nations. Open communications with all groups working in the region. Work to establish equity.
- Help nations stabilize their cultural environments, as well as their populations and resources, which can be related to the limits of place.

The GU would be responsible for defining the goals that are common to human cultures and coordinating goals on the global level.

Claim Ownership of Global Commons

Take legal possession of all global commons, including deep geology, atmosphere and local space, and the oceans. Determine limits of use as well as fees for use. Assign opportunities for use by nations and corporations.

Collect fees until such time as the Earth has been legally incorporated (see later).

Manage Nations & Global Government
The GU would take responsibility to manage the interrelationships of nations, to try to coordinate trade, reduce conflicts, and foster respect for other nations and cultures.

Create a Charter & Constitution for Global Union. The Charter of a global association has to expand the original charter for the United Nations. It needs a formal constitution that spells out additional responsibilities. The constitution would indicate how the Association would relate to nations, corporations, alliances, and NGOs.

Global organization should be in the form of a heterarchy, a multi-level structure integrated at regional and national levels. Community levels would self-reliant, but linked with other communities into nations and regional alliances, The regional alliances would not be allowed voting rights separate from their components in the GU. The regions would be too large to be manageable. Corporations currently that did not have to have a cultural or territorial base might be given temporary status until they were integrated.

Create New Structure & Branches of Global Union. Because a global association will have more responsibilities than the United Nations, it must create a new structure, with new branches and agencies.

General Assembly (Legislative) for All Nations. The General Assembly would include old and new nations. Conceivably, the number of nations could exceed 3000. The General Assembly would make all laws and rules.

Define Limits of Global Law. Global law would be directed at global interactions relating to nations and corporations. It would also address global resources and problems that follow from the cycles and uses of resources.

Make Laws. The UN made laws against war. The GU must expand those laws and be prepared to enforce them. Laws also need to passed to address other human concerns, such as slavery or internal violence. Laws have to be proposed to control international spread of disease. A whole set of laws is needed to control global resources.

Slavery has been outlawed numerous times in the past 200 years, but it still exists. It appears that only a strong international program can stop it, whether it is sex slavery in the Middle East, Japan, and Europe, or open slavery in West Africa, or other forms in Asia, the US and South America.

Resources have to be apportioned, limited or denied by law. Where they are biological, laws have to protect them while they are being restored and renewed, and then for moderate exploitation afterwards. Some resources, such as oil, must have 80% of their reserves left in the ground, if we intend to reduce our carbon burning to reduce atmospheric heating. Other resources, especially rare metals, have to be used sparingly and recycled attentively.

Establish Offices of Normalization & Nationalization
The GU has to issue its challenge to allow cultures or provinces join the organization and to have one vote. Any culture or province would have the right to apply for membership. The size and form of nations would begin to decentralize and normalize. The requirements for nationhood would be basic: A traditional culture, traditional territory, or the size and uniformity of people. Although new nations would vote independently, they could maintain a regional association with other small nations.

Office of Standards for Ecosystems Nations & Trade. Through a eutopian program the Global Union (GU) could set up and enforce international ecological rules of trade, for including environmental costs into the prices of things. Economists, especially global economists, have not much considered these 'externalities' which are really free internalities, free so far, because they have been big and plentiful. To compete in an open arena with fixed economic policies and no ecological policies, nations often squander their ecological capital to be competitive in the short-term. Previously, there has been no reason to save resources to be sustainable in the long-run.

Office of Budget: Secure & Balance Budget. This office would be concerned with funding the GU. Where will money come from? What do we have now? Footloose food and crazed capital? In the form of dues, contributions and the sales of bumperstickers? The new office would have to coordinate the income from taxes on global resources and rents and fees on common lands, such as Antarctica.

Income for International Framework
Right now, the income from member nations is related to their GNP, a partial indication of relative gross national product. With the capability of taxation, the GU would not rely on voluntary national gifts. Although the GU would tax global resources, such as land and air, it could also tax weapons quite heavily. In fact, Ervin Laszlo suggests an international body could tax military expenditures of nations instead of the simple GNP. This would ensure that nations that produced the most weapons would pay the most for the effects of those weapons. Income from gifts and charity could still be used to combat special problems.

Rent for International Framework. The GU could charge rent or fees for global lands and oceans. Ownership of much of the planet would be through a corporation or department. The use of these areas or resources would be rented for the long-term. But, they would also be monitored by the GU.

Membership Dues for International Framework. For the UN, member nations now have to pay a percentage or a flat rate. The United States has the maximum assessed contribution to the UN regular budget—22%. In 2005 the assessed amount is $439,611,612. Actual U.S. contributions to the UN in 2005 totaled $1,959,053,000. This included the regular budget, peacekeeping operations, international tribunals, specialized agencies and subsidiary organizations. The minimum assessed contribution is 0.001%. The scale of assessments for each UN member for the required contributions to the regular budget is determined every 3 years on the basis of Gross National Product (GNP).

Only nine countries (starting with the largest contributor: United States, Japan, Germany, United Kingdom, France, Italy, Canada, Spain, China) contribute 75% of the entire regular budget. Cuba contributes 0.043% of the regular budget. One of the richest countries, Saudi Arabia, contributes 0.713 percent. The size of contributions, however, does not seem to be related to need, importance or dominance.

In addition to their contributions to the UN regular budget, member states contribute to the peacekeeping operations budget and the cost of international courts and tribunals. The level of these contributions is based on their assessed contributions to the regular budget plus variations which take account of permanent membership on the Security Council. UN members also make voluntary contributions to UN specialized agencies and subsidiary organizations. The administrative costs of such bodies, though, are met from the regular budget.

The UN could acquire much more money if the assessment were related to the defense budget of a country instead of its GNP. This would reduce military research and spending. The U.S. defense budget, for instance, was $343 billion a year, almost 200 times its UN membership fee. That kind of assessment would discourage military build-up and perhaps reduce military spending.

Membership dues would be phased out over five years, until tax, rent and other income programs are operating.

Tax Global Things
This term in its most extended sense includes all contributions imposed by the global association upon nations, cultures, or individuals for the service of the association, by whatever name they are called: Tribute, tithe, talliage, impost, duty, gabel, custom, subsidy, aid, supply, excise, or other. Taxes are any charge of money or property imposed by the association upon individuals or entities that are within the authority to assess such charges. Most modern taxes are levied on the basis of economic measurements such as income, consumption, property, and wealth. These taxes would be uniquely applied to global things or processes or to nations.

Use Taxes on Global Elements Cycles. Use taxes would have the effect of limiting the use of nonrenewable resources, such as coal or oil, or the use of slowly renewables resources, such as forest products or fish, located in global commons, such as the atmosphere or oceans. This would discourage overexploitation of previously 'free' resources.

Loss Taxes on Global Resources. The loss tax is on things or processes that interfere with other things and processes, things that cause runaway feedback or the destruction of cycles, things in other words that reduce our continued use of and enjoyment of the earth. This tax would have the effect of internalizing both ecological and social costs; since all consumers would be paying the real costs, no consumers would be protected. It should have the effect of reducing pollution. The purpose of this tax is to change behavior that depletes resources and discourages labor. It also can pay for the damage caused by misplaced or misused resources, such as the destruction of ozone.

Adjustment Taxes on Global Sin & Pollution. There would be no necessity for sin taxes, unless they were necessary to support explicit GU institutions on health, or for such activities in International areas. International pollution would be taxed, however. Especially industrial and agricultural pollution.

Distribution Taxes on Global Things. Distribution taxes have the function of reapportioning wealth. These taxes would be assessed in international waters and lands. They would be on luxuries or incomes that have evaded national and community taxes.

Licensing for a Global Association. A license is a formal authorization by law to do something, such as marry, hunt, or practice medicine. The GU may issue international licenses for some purposes.

Licensing International Corporations. Corporations that have no land base, or that have evaded local charters, would be subject to a license.

Licensing Satellites & Media. Satellites, carriers (or spacecraft), and any items in sublunar or solar system space would be licensed by the GU. These fees would be used for tracking and cleanup. Licenses would be required for wave communications in international areas. Due to the global aspect of waves, it might be necessary to coordinate all media among nations.

Licensing National Weapons. All weapons of a certain kind, from automatic weapons to cruise missiles, would be licensed by the GU. These weapons would be for national police forces and the police force of the GU. Their possession or use would be banned for individuals.

Other Licenses. The GU might offer international drivers licenses. Licenses would be required also for collecting in international places.

Fees & Tolls for a Global Association. Fees would be charged to save, restore, and maintain the common wealth of humanity, such as historical places or unique ecosystems within the global commons. Such fees would be charged to nations.

Global Fees on Common Human Wealth & Great Art. Although much of the common heritage of the humanity on the planet would be free to observe or visit, fees would be charged to limit access. Fees would also be used for special costs such as cleanup. This fee might be applied to special areas of the planet or for works of art that are considered the heritage of humanity, such as many cave paintings or desert designs.

Global Fees on Wilderness. Global wilderness areas, such as Antarctica or the North Pacific Sea, would have limited access. The fees would also support research in these areas.

Labor for Global Projects. The GU could use scientists from other countries. It would also use volunteers from many nations to work on international projects. It could have a formal two-year volunteer program, similar to the programs of nations, but without conscription.

Payout for a Global Association

At this level all costs, including environmental and social, have been internalized. The GU has to be able to guarantee that all unintended costs are being paid, as a result of the economic activities of nations.

Operate Organization & Agencies

The GU has as large a bureaucracy as any national government, in fact, larger than the largest national government. However, since there is an overlap of responsibilities and so many things are duplicated on lower levels, the GU will be essentially a coordinator. Nevertheless, it will have a large number of agencies, as well as the separate governing functions.

Health Supplement. Traditionally, an international body has been concerned with the levels of health of the people of all nations. One function is to make sure that the lower levels of health are improving faster.

Education Supplement. Traditionally, an international body has also been concerned with equalizing educational opportunities for people in every nation. The GU would be responsible to providing education on the common heritage of humanity and on the global environment and history.

Global Planning. Planning is a large part of a global order. The GU must plan for many contingencies that nations may not have to face independently.

Global Resource Surveying & Monitoring. Monitoring is crucial for nations and a global order. No coordinated effort has ever been made to inventory, assess and monitor the ecological systems of the planet. The GU would.

Restore Global Cycles or Places. Many places have been impacted by the international problems of pollution and simplification of land, and disruption of global ecological cycles. The GU would develop programs for restoration.

Create Set-aside Accounts for Global Catastrophes. A large part of the finances and efforts of a global association have to be concerned with global changes in climate and geology. These sometimes catastrophic changes are an integral part of the change and aging of the planet. They have to be anticipated, prepared for, and responded to.

The increase in atmospheric and oceanic heating, as a result of human activities, such as rampant building with concrete, burning coal for heating and even for powering electric cars, and using oil products for other transportation, has to be addressed through cost-intensive projects ranging from ecosystem restoration to the redesign of energy use.

Executive Branch of Global Association
The executive branch of the global association makes sure that the global laws of the GU are obeyed. Other functions of the branch are: To execute policies; to control policies; to appoint officials; to command the police force; and, to veto legislation. The executive branch has to be formed from the assembly of national representatives. It has to be relatively large, and it has to have salaries, support monies, and travel monies. In addition to global laws, the Executive Branch would have to oversee the enforcement of laws by nations, especially where the two levels overlap.

Coordinators of Global Association. The Secretary, equivalent to the former Secretary-General of the UN, would be the head of the Executive Branch. The Secretary would be elected from the General Assembly and serve a term of six years. The Secretary help from the Assistant Secretary, Police Chief, department heads, and heads of independent agencies. The Assistant Secretary would be head of the legislature and next in line to head the branch. The Police Chief would be head of all GU police forces. The Department heads would advise the Secretary on issues and help carry out policies. Independent Agencies would help carry out policy or provide special services.

The executive branch could be headed by an Executive Council of the department and agency heads; these offices would elect the Assistant Secretary.

Security Council of Global Association. The Security Council would be composed of twelve people elected by the General Assembly every three years. There would be no permanent members. The Council would elect its own Secretary to speak for the Council. The veto power would be replaced with consensus.

Assess Threats (Internal & External)
The Security council is concerned with any kind of threat to the global order and the orders of nations and nature. Many of these threats are internal, as nations enter into conflicts with other nations. These threats can be handled by delegations or police actions. But, many threats are external to the planet. They come from solar or interstellar space and tend to be physical.

International Conflict. Widespread poverty may cause catastrophes. Richer countries will need to recognize that the poverty of others is not in their interest, especially as potential markets. Inequity may never be erased. Perhaps some inequity is good and stimulating, but gross inequity needs to be limited.

Each culture develops rules for living together. A common culture provides an ideal framework for public and private decision making. The Sami in northern Scandinavia have institutionalized ways of avoiding conflict, for instance, by shaming those who would impose their will. The Fipa of Tanzania use cooperative exchange rather than competition to keep the peace. The Akawaio of Guyana believe that community disharmony upsets the spirit world, resulting in illness and misfortune.

Conflicts, territorial or symbolic, are symptoms of insecurity. Many of our wasteful conflicts could be more easily resolved through a neutral international power. Conflicts would still occur. Conflicts over prestige or power, as much as for various crusades as for a true state, still lead to human and environmental destruction. The Security Council would be charged with the responsibility to avoid massively destructive forms of conflict, such as biological war or nuclear war.

Global Threats. The planet is still a very active planet. Some threats to humanity rise from the normal activities of the planet, such as volcano-building or continental drift.

Geological Threats. Typical geological threats include earthquakes volcanoes, and mudslides; rare threats may be comets or sunspot activity. While technology may be able to retard some of these threats or counter them early, the best solutions are design. Cities can be moved from floodplains; cities can be required to have quake-resistant buildings.

Climatic Threats. Someone said that civilizations exist through the consent of geology; however, the consent of the 'first empire of climate' might also be needed. Climate is a constant and immediate challenge. Many variables affect climate. One variation, with a long cycle of perhaps 100 million years, is continental drift. When Panama closed two continents, it forced the gulf stream north. The earth's orbit around the sun, a 100,000 cycle, also changes climate. This is close to the spacing of ice ages. Other smaller cycles, at 10,000 years or 6,100 years are minor harmonics perhaps. Other activities that influence climate are sunspots, comets, and volcanoes. The most stable periods seem to be the coldest or warmest weathers. For instance, 400,000 years ago, a warm period lasted 25,000 years. If we are in a 10,000-year warm period, it may be almost over (or last that long again).

Cooling can lead to disease and depopulation, which can lead to further cooling. Farmland is abandoned to trees in times of collapse. Because trees take CO_2 out of the atmosphere, the regrowth of forests after the plagues in 1322-1351 in Europe and China, would have allowed drops in CO_2, which would have allowed the climate to cool.

The years 1997 to 2004 broke most heat and storm records, especially as some of the warmest years on record. Many locations experienced 500-year floods, droughts of the century, and other extremes, within 5-year periods. Some of these events have been related to greenhouse effects.

Oceanic Threats. Ocean currents affect not only islands and coasts, but they affect the entire climates of continents. For example, the effects of the Pacific El Nino current can be linked to droughts in India and China in the past 150 years that caused three times the numbers of deaths from the Black Death, and more than the 60 million who died in WWII.

There does not seem to be a single trigger for El Nino. One trigger has to do with water overflow and then back flow from the western Pacific. Others have to do with sunspots, which would reduce radiation, or volcanic eruptions. Many of these atmospheric, oceanic and geophysical triggers may converge. Nineteenth-century famines may be correlated with ENSO events that influenced China, Indonesia, Brazil, and southeast Africa.

The adherence to political colonialization and 'free market' economics, along with climate changes, made the suffering in famines worse. Millions died within the market system of the golden age of liberal capitalism. Thus, the famines were political and economic failures. At the height of the Irish potato famine, Britain continued to export potatoes from Ireland to Britain. Even at the height of the famines in 1877-78 and the 1890s, Britain continued to export grain from India to England, which had its own agricultural downturn. Millions starved in those two host countries. The invisible hand did not lift those starving in India or China; there was starvation before the British, especially during El Nino events in 1596 and 1630, but many droughts did not result in such widespread and deep famines. Market economics and politics magnified the effects. The market economy can spread risk, through insurance companies, when crises are local and intermittent, but they may not be able to respond when the crises are global and ubiquitous. As the risks increase everywhere, fewer things can be done.

Other movements of water, such as tsunamis or floods, can be linked to earth movements, such as landslides or earthquakes.

Solar System Threats. Solar energy is not quite constant; over millennia it has been increasing; over the next billion years it will start decreasing. Collisions with asteroids or comets will always be a threat due to the nature of the solar system. The passage of the solar system through dust clouds will effect the climate of the planet and the solar output. The GU will have to deal with asteroids and comets.

Provide Security using a Police Force
Countries scramble to identify and claim resources that they need, fighting for them if necessary. Resources in short supply include water, timber, and fossil fuels. Countries strive for resource security, but this leads to further fighting and instability. Traditional ownership is stretched, leading to new disputes. All of these conflicts are the kind that could be resolved by the GU. Cooperative solutions are more durable and effective. Violence only leads to resentment and further violence. The GU has to have either the most weapons in the largest police force or the best moral stance.

Expand Police Force. Peacekeeping, as defined by the Association of Nations, is a way to help countries torn by conflict create conditions for a sustainable situation of concord. GU peacekeepers—police, rather than soldiers and military officers—civilian police officers and civilian personnel from many countries would monitor and observe peace processes that emerge in post-conflict situations and assist ex-combatants in implementing the peace agreements they have signed. Such assistance comes in many forms, including confidence-building measures, power-sharing arrangements, electoral support, strengthening the rule of law, and economic and social development. All operations must include the resolution of conflicts through the use of force to be considered valid under the charter of the Association of Nations.

Use Police Force. The Charter of the Association of Nations would give the GU Security Council the power and responsibility to take collective action to maintain international peace and security. For this reason, the international community should look to the Security Council to authorize peacekeeping operations. Most of these operations would established and implemented by the GU itself with police serving under GU operational command. In other cases, where direct GU involvement is not considered appropriate or feasible, the Council would authorize regional organizations such as the North Atlantic Treaty Organization, the Economic Community of West African States or coalitions of willing countries to implement certain peacekeeping or peace enforcement functions. In modern times, peacekeeping operations have evolved into many different functions, including diplomatic relations with other countries, international bodies of justice, such as the International Criminal Court, and eliminating problems, such as land mines, that can lead to new incidents of suffering or fighting.

The GU would be expected to have many operations around the world, similar to the peace operations of the UN. Recent operations of the UN in Africa include the Burundi Civil War (2004), the Civil war in Côte d'Ivoire (2004), Second Congo War (1999, United Nations Organization Mission in the Democratic Republic of the Congo, Second Liberian Civil War (2003), the Eritrean-Ethiopian War (2000), the North/South Civil War and Darfur conflict (2005), and the Moroccan occupation of Western Sahara. In the Americas, the UN monitored the 2004 Haiti rebellion. In Asia, the

UN was involved in: The 1949 Indo-Pakistani Wars. In Europe, the UN put itself in: The 1964 Cyprus dispute, the 1993 Abkhazian War, and the 1999 Kosovo War. In the Middle East, the UN participated in the 1974 Agreed withdrawal by Syrian and Israeli forces following the Yom Kippur War, the 1978 withdrawal of Israeli forces from Lebanon, and the 1948 various cease-fires and assists. All with varying degrees of success.

Deal with International Conflicts & War. The police force would be entrusted to try nonviolence as a first response. In some cases police would be unarmed. If not that did not work, police would use appropriate weapons.

Conflict cannot be separated from other things, such as environmental destruction or inequities. For that reason, the Security Council and its force has to consider the broadest meaning of security: It is for people to have the resources and opportunities to provide themselves with their needs. Thus, security has to be addressed on many levels. For example, what do forests have to do with peace? Of the fourteen nations requiring UN peace-keeping operations since 1990, twelve of them have lost over 90 percent of their forests. Perhaps peace-keeping should be abetted by tree-planting. As ecosystems are destabilized, nations become less stable, and must be helped with more than social conflicts.

In the cases where conflicts cannot be resolved, perhaps the GU could oversee ritual combats between the heads of nations to decide who was right. With the removal or large-scale weapons and the establishment of a general equity between nations, the remaining issues may not excite the same violent passions that result in organized slaughter. We have to perfect the art of resolving conflict. Mastering it through social debate would free resources to satisfy basic immediate needs of food and water.

Protect against Interference Events or Disasters. Police would be expected to be prepared for natural disasters, including the extraterrestrial. The police would focus on the prevention of man-made disasters, from watersheds that were compromised to chemical spills. Collisions with asteroids will always be a threat due to the nature of the solar system.

The flexibility of a natural ecosystem to respond to change has been reduced, so either we have to take over control, which could cost quite a lot, or we have to adapt to the diminishment after some climatic episode.

Agencies for a Global Association
Agencies of the Association of Nations are necessary organs to present information and research to the main branches of the Association. A partial list follows.

Food & Shelter Agency. This agency is devoted to monitoring the food and shelter requirements and deficiencies of groups in every nation. It would provide information for traditional or ecological ways to build food supplies or shelters. Through international volunteers, it may provide help with expertise and labor.

All Energy Agency. This energy agency would consider all possibilities: Human energy, animals, fire or combustion, fossil fuel, solar energy, geothermal energy, wind energy, and nuclear energy from fission or fusion. It may recommend and sponsor less hazardous kinds of energy use, such as solar energy and wind energy. It would also work to make sure that energy is not wasted; this may mean building or converting more efficient houses. Without the conservation of energy, giant wind farms or solar farms will dominate the landscape, ruin the scale and be possibly as hazardous as fossil fuels or nuclear power.

Transportation Agency. Transportation is a global phenomenon. In addition to land, many kinds of transportation use the atmosphere and oceans—in fact many new proposals have to do with the undersea environment. The GU would monitor or control traffic in the global commons, especially related to regional highways and roads into wild areas.

Economics Support Agency. Economies are tied together now in global patterns. They might be overconnected. This agency would provide information on economic weaknesses and problems to nations. It would act to provide alternatives to some kinds of economic practices and suggest practices that would eliminate the worst excesses.

Education Support Agency. Education needs to be stressed. The goal of education is for people to choose within limits—limits on wealth, waste, and freedoms that might endanger others. Education allows people to choose without being conditioned by brainwashing or dishonest advertising. The UN supported programs to enhance education, and it established the United Nations University for Peace, Costa Rica (a nation that has abolished its army), as an institution of higher learning for education for peace. The GU would continue these kind of efforts.

Health Agency. Health needs to be stressed, in every nation. The GU needs to monitor global disease and threats, to reduce chronic disease. Chronic diseases are neglected conditions. Chronic diseases represent a huge proportion of human illness. They include cardiovascular disease (30% of projected total worldwide deaths in 2005), cancer (13%), chronic respiratory diseases (7%), and diabetes (2%). Two risk factors underlying these conditions are key to any population-wide strategy of control: Tobacco use and obesity. These risks and the diseases they engender are not the exclusive preserve of rich nations. Chronic diseases are a larger problem in low-income settings. Research into chronic diseases in resource-poor nations indicates that it is critical to intervene early in the course of any epidemic. Fast intervention could save many millions of lives.

Communications & Technology Agency. There is a need for the development and delivery of a strategy for schools and other learning and skills institutions. This agency could provide strategic leadership in the innovative and effective use of communications and technology to enable the transformation of learning, teaching and educational organizations for the benefit of every learner.

The agency would be charged with regulating interstate and international communications by radio, television, wire, satellite, and cable. Its jurisdiction would cover in fact every nation and international areas.

The development and application of new technologies needs oversight. Technology is changing faster than most cultures can adapt to it. New technology is constantly altering our bodies and brains, as well as ecosystems and material flows. But, there are always losses to go with gains, and unintended consequences for every change; these have to be identified and dealt with. Changing humans genetically for instance, involves millions of genes in thousands of contexts. Improving people in a uniform holistic way will involve tracing connections backwards and forwards in space and time. It might work, but we have to be prepared for failure. That should not be a problem; science is a form of play where most experiments fail. And, we have to think about ways to reverse changes or make them less negative.

Research Agency. This Agency is the central research and development organization for the planet. It manages and directs selected basic and applied research and development projects for the international body and member nations. It pursues research and technology where risk and payoff are both very high and where success may provide dramatic advances for issues of global importance.

The reductionist path taken by science has yielded tremendous results about how the world is built up out of particles and molecules. Now that we have uncovered the complexity, we need to address relationships. This is where a synthetic path can help, by identifying emergent principles and operations. Science is an open, self-referential, self-correcting system capable of using analytic and synthetic methods.

Advances in science have been quite remarkable. How can it continue being remarkable, but applicable to whole systems? While pure research continues to reveal unimaginable details of ecosystems, and while applied research continues to support sophisticated use, more ecological and landscape research is needed. Research is expensive, time-consuming, labor-intensive, and uncertain, however.

Long-term research is especially important in forest ecosystems, since many of the components live hundreds and thousands of years, and the forest itself can live far longer. Long-term research requires different levels of monitoring, including environmental, biological, and ecological. Environmental monitoring is an umbrella for many activities, including climatic variables and geological processes; for example, the systematic recording of soil and air temperatures, humidity, air pressure are measured by meteorological organizations to predict long-term climatic change. Long-term research also depends on a stable cultural base and shared values between generations.

Other kinds of research needed equally are: Historical research, fire research, productivity research, mortality patterns, mature ecosystem research, key species, artifact interaction, and genetic research.

Business & Corporations Agency
Businesses and corporations sometimes have more people, money and power than nations. This agency would be concerned with monitoring and assisting these entities, to ensure that they do not violate the norms of nations or take advantage of cultural limits.

Incorporate Earth. Political systems are impotent to stop the massive interference in ecosystems by international corporations. The simplest and most direct way to give the earth a voice in the development of the earth by humanity is to incorporate the earth following international law. The entire planet, with its biochemical cycles and nonhuman communities, would become one legal body. Since corporations are human constructs, however, humans would have to represent ecosystems and their wealth of living organisms.

In early civilizations, the advancement of the state was expected to contribute to the welfare of its people. Corporations are recent devices created by states for public purposes. Most early U.S. corporations, for example, were concerned with travel (turnpikes and inland waterways) or safety (fire insurance) — they resembled public agencies more than profit-seeking associations. The exclusive privileges and political powers granted to corporations were based on the implicit promise of social services.

The association of economic development with national wealth allowed incorporation laws to be broadened. The corporation was given the constitutional rights of an individual, accidentally. A corporation is a legal entity, independent from its founders, with its own rights, privileges, and liabilities. It is, however, required to obey laws and pay taxes; and it is accountable for its deeds in courts of law.

Unfortunately, as private good became identified with public good, corporations became larger, more acquisitive, and less concerned with social services. The quest for profit now has the effect of violating social amenities, such as clean air and clean water, instead of ensuring them. No responsibility is taken for environmental degradation since no right of contract or fair use of property has been breached.

Although current wisdom[15] holds that a corporation's only responsibility is to its stockholders, corporations are being pushed to include social purpose in their strategies, again. Alas, they are doing poorly at it. They do not know how much responsibility to take, or where to put limits, or whether to pursue policies that diminish their profits. Corporations have proved spotty in doing social and environmental good. It would be more appropriate to have them deal with the environment as a corporate entity concerned with maximizing its own values. Of course, that would mean no more 'free' resources or environmental services.

The important advantages to incorporating the earth are the same as for incorporating a business. There are at least six advantages: (1) Managerial flexibility; the stockholders are separate from managers; responsibilities are assigned by needs of the corporation. (2) Limited liability; the corporation borrows and repays. It shields its members from hazards to which they would otherwise be exposed. (3) Financial advantage; the ownership of assets can benefit stockholders and the corporation. (4) Tax advantage; investments in the good of the corporation may not be taxed by nations. (5) Estate planning and longevity; the corporation exists indefinitely beyond the lives of its participants. (6) Central management and representation; a large and complex business needs operational and managerial efficiency. Many of the participants have no direct voice in the operation—they must be represented.

The Earth Incorporated would focus on a core business: To ensure the integrity and continuity of life and all its connections and to secure the opportunity for development free from undue interference. It would operate to optimize values, like any good corporation, but the values would be ecosystem values, such as fungus values and earthworm values, as well as human values.

A temporary Board of Directors would adopt bylaws, elect working officers, approve stock certificates, open accounts, and arrange a stockholders meeting. The stockholders would elect new directors, possibly from Association of Nations representatives or directly from elections, and decide on dividend declarations.

Stockholders, as citizens of independent nations, would turn over common and national property to the Earth Corporation, which would issue stock certificates to the stockholders. The corporation would allocate the purchase price of stock to capital at par value. Most of the shares—the percentage to be determined by the board as necessary to the operation of ecosystems—would be treasury shares. Anything more than par value would go to capital surplus, and only capital surplus could be distributed as dividends. Stockholders have the right to receive these dividends equitably, without resort to traditional distributions of wealth.

Stock certificates denote ownership of the corporation. Although the stockholders own the corporation, they do not own the property of the corporation, the earth, which is owned by the corporation itself. Stockholders, as individuals, groups, or nations, could make agreements about how business would be conducted, about what resources would be used or traded.

The elected board of directors would make decisions of distribution and limitation. Percentages would be deducted from the interest for the operation of the corporation and for equitable distribution to nations less favored by chance with biological or geological wealth. Furthermore, since the dividends would be distributed among people according to net ecosystem productivity and resource availability, no advantage would be gained by nations having large populations.

The basic functioning system would be considered capital, thus limiting the amount of human use of resources and probably the size of human populations. Interest would accrue in the form of net ecosystem productivity and diverted percentages of materials, such as gold or water.

The earth incorporated would solve the problem of having to value ecosystems in monetary or quantifiable terms; its systems would be untouchable capital. The human value of resources like copper, air, or water would be equated to the technological cost of recycling or producing them.

Raw material and energy are only two facets of the capital of a corporation—another is human ingenuity. Thus, human wealth would not be limited by restrictions on the availability of resources, but rather by a shortage of ingenuity.

An incorporated earth would be instrumental in conditioning international corporations to their social responsibility and in internalizing all costs. This corporation and governments could use traditional means, such as credit access, low interest rates, and setting priorities on equity issues, to evoke public interest in smaller and healthier human endeavors. The corporation would keep rights to global territories and resources, issue leases to nations, and monitor scarce resources.

The suggested articles of incorporation are:

FIRST: The name of the corporation shall be The Earth, Incorporated.

SECOND: The purposes for which the corporation is formed shall include: The protection of functioning ecosystems and their living beings from destructive interference.

The conduct of inquiry into the operation of such systems and the role of humanity therein for scientific and educational purposes.

The taking of appropriate legal steps to carry out these purposes.

The maintenance of all real common property, including all lands, seas, and atmosphere, subject to the restrictions and limitations hereinafter set forth, to use only the interest from income therein, reserving the principal thereof exclusively for the aforesaid purposes, it being intended that the corporation be organized and operated for preservational purposes and not for pecuniary profit.

The corporation is organized as a voice for nonhuman beings and systems. No part of the income of the corporation, if any, shall inure to the benefit of any trustee or officer of the corporation or to any private individual having an interest in the corporation (except for reasonable compensation) and no trustee or officer of the corporation or any private individual shall be entitled to share in the distribution of any of the assets of the corporation.

The corporation shall not be authorized to carry on propaganda, influence legislation, participate in any political campaigns, or discriminate against human cultures. In furtherance of the foregoing purposes, the corporation shall have the following powers:

To accept and hold by gift or judicial order any real or personal property of whatever kind, nature, or description, wherever situated.

To sell, transfer, or dispose of the interest from any such property, but not the principal or any part thereof.

To make, accept, endorse, execute, and issue bonds, promissory notes, bills of exchange, and other obligations of the corporation for monies borrowed for the purposes of the corporation.

To invest and reinvest its funds in stock, bonds, or in such other securities and property as its trustees shall deem advisable, subject to the limitations and conditions contained in any grant or gift.

In general, and subject to such limitations and conditions as are or may be subscribed by international law, to exercise such other powers which now are or hereafter may be conferred by international law upon a corporation organized for the purposes hereinabove set forth.

THIRD: The operations of the corporation are to be conducted on the surface of the earth but the operations of the corporation shall not be limited to such territory.

FOURTH: The principal office of the corporation is to be located temporarily in the Global Union of Commonwealths (GU), currently in the City of New York, State of New York, United States of America (and possibly relocated to Antarctica or elsewhere).

FIFTH: The number of directors, who shall be known as trustees, of the corporation shall be not less than 30 (a minimum number associated with major ecosystems), nor more than 3,300 (the number of independent cultures associated with biogeographical provinces and subprovinces).

An earth corporation would have a special charter to protect the planet and provide opportunities and services for all the inhabitants.

Recharter International (Landless) Corporations. Nations are people in place, whereas corporations are people in profit, often out of place. Corporations are like the feudal domains that evolved into nations. They are new kinds of political organizations, rich and arrogant, but they are currently free to leave behind any social or economic mess as they search for cheaper labor or freer markets.

Nations need corporations more than corporations need nations, Lester Thurow notes. But, corporations are legal individuals. Nations simply need to make laws requiring that every corporation have a national base, preferably where most of the employees work, and must meet their obligations and duties as national citizens.

Laws for corporations, such as Antitrust laws, need to be consistent globally, says Thurow. Laws, like morals, have to react to new technology and new social situations, regarding consumer protection or company protection.

The GU could make a binding treaty on transnational corporate accountability. The GU would also collect data on Transnational corporations. Local corporations would have national accountability.

The question is who should have the power and how much. The GU would have the power to charter transnational or global corporations. The GU could tie corporations to place and force them to be responsible by identifying the corporation as a legal community in a place, rather than a fictitious artificial individual, with no address. Members, or staff and stockholders, would share equal responsibility and liability for the activities of the real corporate community. The GU would recognize that the corporate community has moral obligations which can be spelled out. The GU would recognize that the community is not permanent or abstract. It must have a place and participate in that place as a responsible community.

The GU would create a comprehensive model of requirements, such as stability, justice, and balance for transnational corporations. It would make sure that the charter requires serving public interests as well as private gains, to increase the meaningful work of employees, who want to serve the public good of the place as well. The GU would make sure that profits and losses, gains and costs are all privatized equally. The GU would make sure that all costs are internalized, that the capital of the ecosystem has to be paid for use. The GU would require fair labor practices in terms of reimbursement, where the minimum wage would be above a standard poverty level for a nation and the maximum could be no more than ten 10 times the minimum.

The GU would require a corporation to address environmental problems, using ecological performance standards, and address social problems, using social performance standards. The GU would require plans showing dependencies on environment and society community, then monitor corporations for their economic health, to make sure that they develop and mature correctly. The GU would also make sure they have plans for unforeseeable contingencies.

The GU would monitor corporate responsibility, the natural home communities of corporations for their ecological health, and the health of the home human communities. The GU would set standards for corporations to internalize the loops of production and waste, at least within the system of sharing corporations. The GU would track all impacts. The GU would encourage corporations to promote ecological design, and to guarantee the use and safety of their products and services.

The GU would expect loyalty from corporations, but try to develop loyalty both ways.

Set Up Trusts for Regions. The form of a nonprofit corporation is not the proper approach to protect regions. Perhaps, a region could be represented well by some sort of legal trust, as are private properties. This might solve the dilemma of ambihuman species as well as future human generations, which require a much longer time frame than most plans. Bali's water-sharing temple system for rice farmers is a good example of commons management. And, there are other examples of this kind of management of limited resources. The Spanish Huerta was also a system for distributing water and resolving disputes.

A regional trust for Bali could have elements of a constructive trust, since it would be imposed by law as an "equitable remedy" against those holding the assets as a matter of luck or discrimination. It would have elements of a spendthrift trust, since humanity is unable to control its spending of natural capital; the trustee, perhaps an agency of the United Nations, would have the power to spend only ecological interest. It would resemble a unit trust, in the sense that all human beings would possess a certain share of the interest of the planet—of course, nations would control a percentage, determined by the ecological carrying capacity, which would be divided equally among the population; this would encourage nations to normalize their populations. It might resemble a public trust, in the sense that it would have the object of keeping the planet healthy, as the source of most capital.

Co-ownership, in a trust, could be divided between species first—that is, the ownership of the home, earth, is shared by all living beings, and all living beings should have some legal representation. For Bali, living water would have an interest. If it were a hybrid trust, the amounts of the trust interest could be paid out at the discretion of the trustee; this might be used to settle long-standing grievances and inequities. It could resemble an incentive trust—it would encourage some behaviors, such as inventiveness and frugality, and discourage others, such as waste or inequity.

Obviously, there would be many benefits to setting Bali up as a trust. The greatest benefit would be legal protection of many areas for ambihuman species. In fact that would be a major purpose of a trust. A second major purpose would be to equalize the income from global interest so that it would be divided equally among residents.

Other regions, such as the Palouse Grasslands or Tall-grass Prairie in the US, the Siberian Tundra, or the African Rainforests, could also be set up as trusts, to be owned by the GU and participating nations.

Office of Personnel & Service. This office keeps track of permanent and temporary employees of the global association. The association would use a civil service examination, much like that developed in China, to qualify and individuals for positions.

Permanent & Part-time Staff. The permanent staff would support the functions of the body. All positions, even those of the assembly, would be paid by the GU organization, with the understanding that the first allegiance professionally would be to the global body, before any national or community interests.

Volunteers Service to GU or Nations. The GU would create a separate volunteer program, for those who wanted to volunteer to work in global areas, such as mid-oceans or Antarctica. Some of the volunteers could serve on peacekeeping or emergency response missions in specific nations. The specific commitment could be two years, with extensions.

Office of Self-Assessment & Future. Self-Assessment in an organizational setting, according to definition, refers to a comprehensive, systematic and regular review of an organization's activities and results referenced against a model organization. The Self-Assessment process allows the organization to discern clearly its strengths and areas in which improvements can be made and culminates in planned improvement actions which are then monitored for progress.

Self-assessment can be extremely valuable in helping an organizations to critique its own activities, and form judgments about its strengths and weaknesses. For obvious reasons, self-assessment is more usually used as part of a formative assessment process, rather than a summative one, where it requires certification by others.

The UN needs to have access to records to assess the performance of independent governments. So that in addition to a self-analysis at a national level, there will be a GU assessment. Assessment could be based on two-year cycles, although planning could extend to 500 or more years.

Judicial Branch & International Courts
The judicial branch of government is made up of the system of courts and offices. The Interests Court would be the highest court on the planet. The Constitution would establish this Court and all other courts. Courts decide arguments about the meaning of laws, how they are applied, and whether they break the rules of the Constitution. The functions of the judicial branch are: To maintain the integrity of the Constitution, to interpret laws, and to check the executive and legislative branches and special commissions, by monitoring their activities.

Monitor Other Branches & All Laws. The judicial branch monitors the Executive Branch, the Legislative Branch, and all other departments and agencies for the relevance of their behavior to the constitution of the organization.

Interpret Laws. The judicial branch interprets the laws made by the legislative branch and their enforcement by the Executive Branch for their closeness to the constitution and the intents of the peoples of all nations.

Conflicts & Interests Court. The Judicial Branch would set up a court to resolve interests and conflicts of the member nations, especially as related to global resources. The Court would decide matters relating to use of global resources or shared resources.

Environment Court. The UN Environmental Programme is relatively weak, from lack of capacity, staffing and funding. It should be made an agency that could coordinate environmental policies for the planet. It should have enforcement powers, maybe taxing powers directly. The environmental agency would conduct environmental surveys and monitoring.

The GU Environmental Court could address all global properties, including the atmosphere, deep continents, oceans, moon, space, bioregions, watersheds, and wild habitats. Many of its issues would be raised by the environmental agency.

Tax Court. The Tax Court would ensure that the taxes of the global association are legal and being collected. It may also ensure that the taxes of nations are not out of line with those of other nations.

Police Enforcement Court. This Court would judge cases related to police actions of the global association, in handling problems between nations or corporations.

Office of Rights (Human & Ambihuman). Should some rights be made universal? We know that there are universal rules for human behavior, most of which relate to families. We also know that some things should be made universal by law.

An economy has traditionally been seen as a morally neutral body, but even if it has only to conform to the nominal rules of society, it is already a moral agent. Economies are no more neutral than other organisms. Many areas of moral concern already are recognized: Worker safety; affirmative action; advertising truth; foreign investments; and harm to the consumer, public, and environment.

Responsibility occurs wherever the interests or rights of a person, society, or ecosystem are significantly affected by the actions of economic actors. Responsibility can be understood in terms of costs and benefits, that is, through operations and their consequences rather than abstract behavior. Every action entails a gain and a cost (or profit and loss). Profits and losses are distributed privately, socially, or environmentally.

Economies need to work cooperatively to make sure the costs and benefits are extended equally throughout the system. They could start by sponsoring the rational use of rare resources through taxation. Influence the government to determine priorities for wilderness areas or special landscapes. Beautiful, fragile, unique, or endangered ecosystems and species must be protected at the expense of commercial activity. Assigning rights to Nonhuman patterns would support many kinds of preservation.

Office of International Trade & Equalization. The resources of the planet are spread unequally around the globe. Local trade allowed some resources to be acquired through trade, especially things used for ceremonial activities, such as ochre or gold. Trade also exposed people to new ideas, and to distant people and their products. Trade also forced the collection of surplus in excess of needs and perhaps in excess of the system. Many effects of trade are long-term problems and do not become evident for several generations. They are also very difficult to reverse. For a society that needs surpluses to continue, with growing dependents and growing numbers of people, there is little flexibility to change. Other cultures, e.g., the English, encouraged trade to increase wealth.

Regional trade started to emerge a few thousand years ago, with Roman trade between 100 BC and 400 AD and Mongol Trade with China and Europe, 1250-1350. The Atlantic slave trade in the 1700-1800s linked three continents. The first wave of global trade destroyed many of the traditional societies of the Americas. A later wave of globalization destroyed traditional economic systems. European exports, especially in textiles, undermined regional livelihoods. These processes enriched the Atlantic shores. It also widened the inequities within nations.

Economics has become more and more global. Where peoples used to trade material goods, fish for roots or feathers for leather, for instance, now all things have a common symbolic value, most often expressed in yen or dollars. This means that whoever works the cheapest sells the most.

The purpose of this office is to promote the equalization of trade opportunities. It has to make sure that trade partners are not overconnected. The office would also help nations that are underconnected by trade. The office would safeguard the global aspects of the system. It would create rules for the international regulation of trade.

Globalization has to be managed to protect the environments and the workers in a nation. The globalization of economics, at least the implementation, leaves many things out of the equation: Local decisions, importance of representation of labor, the fragility of local ecosystems. The health of the local economy depends on local agreements, but now those can be repudiated by global agreements on global integration. That is a poor trade for workers, in the name of global free trade. Economic globalization is premature and dangerous without a political framework at the international level to moderate the unfair advantages from trade that is not free of profit-focused corporate or political decisions. There need to be international rules and laws to slow the acceleration of trade for profit. They need to be expanded and reestablished before global trade is beneficial to everyone, as it could be.

Commission on the Earth
The Commission would have trusteeship over global commons. Boundary issues need to be resolved first. The Commission would limit national jurisdiction of oceans to 20 miles from the border, rather than 50 or 200 miles, to protect ocean resources. Furthermore, it would have a say on continental shelves and shore fisheries. The other 65-70 percent of the surface of the planet has to be regulated as a common area.

Environmental Surveys & Monitoring. Surveys need to be made of every kind of ecosystem. Humanity needs to have an inventory of kinds of systems and kinds of changes. How is conserving terrestrial animals part of conserving ecosystem health? We do not know whether animal declines were caused by disease or some other factor, such as competition or predation. We need to find that out.

Monitoring is the key to understanding changes. Disease needs to be monitored as an important indicator of integrity. Other indicators are surveys of key species, habitat mapping and human impacts monitoring. Complex interactions have to be monitored, using a range of indicators at levels from behavioral to ecological. There may be limitations of the bioindicators of ecosystem health. Perhaps we need to find common and endangered indigenous species and monitor them, hoping that would reflect the health of the entire system.

Create Inventories. The Commission would inventory every ecosystem on the globe. An inventory is a complete list of all components of an ecosystem, from the geological to the ecological. A complete inventory of elements starts with the shapes of the features in the area, the geomorphology. The large volumes are rounded and natural hills — even the agricultural evidence is almost natural, that is, from the roadside not the air, the fields appear not to be squares, triangles, or circles; a small number of geometric shapes exist in the buildings by the road, but because of their scale are not too intrusive. A complete physical and ecological inventory would then be integrated with economic and cultural values.

With the information available now, from a more extended resource inventory and with optimal ideas about renewal, climatic conditions, traditional land-use patterns, local cycles, and ecological requirements (limits), conservation is more effective. Its goal is to support a steady state economy within optimum ranges based on natural and human limits.

Create Monitoring of Global Properties. The Commission would address real global problems, such as global warming, which has resulted in grain harvest shortfalls in recent years. The climate in general would receive renewed and detailed examination.

A global system is the sum of localities and may have unique characteristics of its own, that is, the universe has characteristics that local frames of reference do not have. An ecosystem is directly connected with global cycles and other ecosystems. The system is embedded in larger systems and global cycles, cyclicity. There has to be a good substrate with energy and materials flowing into the system.

The global problems include: Global warming; ozone depletion (chemical caused); disruption of global cycles; contaminations (nitrates, mercury). These threats cause ecosystem collapse. Ecosystem breakdown happens as a result of stresses, singly or grouped, that relate to interference patterns in the system, most of which are caused by the human species now, although the potential for asteroids or volcanic eruptions remains.

Our actions on the planet are experiments, whether we want them to be or not. Ignorance, denial, or cupidity cannot unmake this experimental course, which may be global and irreversible. This Commission would make the experiment conscious and cautious.

Atmosphere Monitoring. The biosphere can exert control on the temperature of the surface and the composition of the atmosphere. On the other hand, soil types and the weather can limit vegetation; invasions of vegetation change soil types. An ecosystem is a topographic unit, a volume of land, occupied by organic beings, extended over an area and through time, with connections to larger mineral, chemical, water and air cycles. This means they are geographical units that intersect with atmospheric units. Ecosystems have a vertical structure, that includes the levels of climate from micro to topoclimate and macroclimate, to soils, water structures, and bedrock, as well as a horizontal structure. An ecosystem is a process, or a set of interlinked, differentially-scaled processes that may be diffuse in space but are easily defined in turnover times.

Processes encounter each other in a functioning web of an ecosystem, with tangible and diffuse surfaces. Lynn Margulis qualifies her definition of an ecosystem: The smallest unit capable of recycling the elements of its membership. For example, organic carbon can be expired, fixed, reacted, or transformed. This is done through the physiological activities of the members of the system, through breathing, enzymes, or some other way. Margulis states that elements recycle faster within ecosystems than between them. Forests, for instance, act as sinks for carbon. The rapid release of sinks can affect other atmospheric or terrestrial cycles. The biota of the planet appear to regulate the surface temperature, atmospheric composition, and ocean chemistry, for a start, perhaps like the human body regulates its temperature, blood chemistry, and other vital signs. As it achieves a new balance, with human inputs, the atmosphere may cause problems with agriculture and other human activities. Atmospheric monitoring is tied with water and terrestrial monitoring. Parts of the atmosphere could be designated as reserves.

Deep Continents Monitoring. Continents are formed by the movement of tectonic plates. As each continent forms, it develops its own combination of topology and water and climate patterns. Australia for instance has become the driest continent now. As continents rise or subside, in addition to their movement and combination, life has to adapt to the changes. The purpose of such monitoring would be to predict long-term changes as a result of continental change.

Bioregions Watersheds & Habitats Monitoring. Monitoring bioregions and watersheds involves a larger spatial scale than ecosystems. Regional goals are appropriate for bioregions. Evaluation of data must occur in an integrated manner that spans biological and physical scales, watersheds, administrative boundaries, as well as functional areas. To understand how ecological processes are connected we need to relate information across disciplines and agencies, and collectively perceive the effect of our actions on the environment. This approach follows ecosystem theory (the hypothesis that cycles in nature integrate the physical, chemical, and biological components of ecosystems), and the hierarchical organization of ecosystem functions throughout the landscape. Hierarchy theory can be described as the development and organization of landscape patterns, e.g., vegetation communities, through time and space.

This can be accomplished by incorporating the three primary attributes of biodiversity, as described by Jerry Franklin—composition, structure and function—into four levels of organization—province, subprovince, watershed, and site.[16] Indicators incorporating composition, structure and function at the appropriate levels of organization have been identified for many ecosystems; they range from landscape morphology to human demographics and cultural influences.

For watersheds and habitats one has to consider the impact of any kind of vegetation removal. What is a minimum, optimum or maximum vegetative cover for various watersheds? Science might identify minima or maxima but philosophy and conservation can aim at optima.

Antarctica Moon & Space Monitoring. The moon has been such a constant for the earth, it might be hard to imagine how things would develop or change if the moon changed or were destroyed. The moon is related to stability of earth system. Because of its relatively large size and closeness, the moon forms the other half of a double planet with the earth. The moon revolves around the earth and both orbit the sun, so the entire lunar cycle takes almost 30 days. The moon exerts a gravitational pull on the earth that is stronger on the closer side. This creates a tidal variation in the heights of the oceans; these vary monthly. For many shallow water creatures, amphibians and mammals, it is good to adapt to these tidal variations. Of course, the earth exerts a pull on the moon, also, but it is less dramatic.

Due to its rotation around the sun, and to imperfections in balance, the earth tilts on its axis. This obliquity of the ecliptic creates seasonal variations, to which most animals and plants have adapted. Any changes, even relatively small ones, could be catastrophic for climate—a one-degree change could account for some ice ages. Jacques Laskar and others have documented the importance of the moon on the habitats of the earth. A stable climate needs the influence of the moon; otherwise, there would be immense variations in the solar heating of the earth's surface. The moon provides energy pulses, stabilizes axial tilt, and causes tides and variations.

Space is an equally important part of the system environment. It is not only the source of the sun, but it is the sink for energy from the earth as well. Stars and the sink of space provide many elements and their proportions. Galactic and solar system dust influences long cycles of the earth's climate.

Life is also challenged by energy and gravity, as well as the moon's behavior. The moon provides daily variations in tides, that provide energy to organisms, although the organisms have to adjust for the different levels of water.

Office of Ocean Commons. Life had once been limited to the oceans. The evolution of living forms expanded those limits. Life, over time, has colonized deep oceanic vents, as well as Antarctic gravel fields. Oceans are not as productive as most land-based ecosystems. Worldwide, oceans could only support about twenty two million people, even though its area is over twenty three times that of grasslands. The ocean bottom acts as a sink for phosphorus. The rapid release of sinks can affect other atmospheric or terrestrial cycles.

This Office would protect the integrity of the oceans. Right now, probably over 66% of fish stocks are overexploited on open ocean (maybe over 75% in areas under national control). Predatory species such as tuna and marlin have fallen 90% in 50 years. Possibly 50 species are driven extinct every year. Over 50% of coral reefs are damaged by mining and warmer water temperatures (perhaps 20% have been destroyed). Algae are producing less chlorophyll and less oxygen; concentrations have fallen 9-12% in under 10 years. Ocean fishing and fish mining is destroying the very support systems. Aquaculture is creating more problems with pollution, diseases and parasites. The dumping of trash, especially plastics, and oil products is killing entire habitats and creating dead zones everywhere.

This situation is so extreme that the UN will have to create and enforce immediately a series of laws and rules. First, it must create a model of the historical abundance of the ocean, along with a program to inventory the current state of the ocean. Another program will be established for monitoring the ocean. These programs will determine the rate of exploitation after any recovery in 20-60 years (and it may be only 40-50% of the highest rates decades ago).

Then, ocean fishing will have to be decreased by as much as 80-90% immediately, allowing only some traditional or low-impact modern systems. Whale and dolphin harvests will be suspended. This will not be popular with many groups and corporations, but it is necessary if stocks are going to recover and can be reasonably exploited in 2 or 3 human generations; otherwise the oceans are going to start collapsing within one generation.

Aquaculture, shoreline and open ocean, will have to be suspended until it can be regulated and monitored properly.

The dumping of plastics will have to be discontinued; great efforts will have to be made to clean up the plastics in the great ocean gyres covering 40% of the ocean surface (this could be done with paid jobs or volunteers through the office). It will require new technology.

Large marine reserves, with the status similar to national parks on land, need to be created for 50-75% of the ocean surface. Some would be designated ocean wilderness or conservation areas. Some reefs may have to be restored.

A special program should be started to restore algae and fish populations, especially large predatory fish like sharks, tuna, swordfish, and whales. Because the oceans play such a large role in regulating climate, they need to be healthy; this means that we need to rewild the ocean by letting sperm whale numbers return to over 120,00, Blue whales to over a million, and the other great whales to their precatch numbers. Sharks, tuna, and swordfish need to be allowed to recover in their millions (we do not even know how many there were before 1900, only that they are all becoming endangered from hunting). These animals all fix carbon dioxide (easily over $13 billion in carbon trading per year). *We need sharks and whales (and fish and plankton and —) for the oceans to become healthy again!*

The cost of lost income might be $60 billion per year; the costs for restoration may approach $10 billion per year, but the losses from a sick or dying ocean are incalculable. A healthy ocean might feed a billion people in the future. The aesthetic value would translate into more income from tourism for bordering nations.

The UN would create and enforce a mandatory global registry of all ships; some factory ships would have to be retired or repurposed.

Finally, the UN would encourage national governments to end subsidies for fishing. Mismanagement causes a vicious cycle where fishermen race to catch the remainders of species and drive the population below any recovery. Also, nations need to tighten national rules and make sure there are mechanisms for enforcement. Because the threatening problems of acidification, erosion and pollution originate on land, nations need to control those. Nations will have to ensure that their rivers provide a healthy flow into the seas and ocean. The UN could reduce the 200-mile limits back down to 10 to solve many problems.

We may have to change our eating habits, or go hungry for a while, but if we do not stop eating the ocean alive now, there will be no choice later. We can work to conserve the ocean and let it repair itself and return to health, but only if we follow these rules until it does recover.

Ecological Design & Planning Commission
The Commission is entrusted with finding out what is on the earth, as well as designing forms for the continuity and enhancement of life on the earth.

Create Long-term Ecological Plans. Despite valid arguments against centralized and global planning, the most important thing people can do with civilizations and ecosystems is plan. That is, plan for the ecosystem needs and for human needs, plan for landscape, watershed, preservation, site, alternative use, and social objectives.

The goal of the Commission is to create a practical plan that fits cultures into nature in ways that protect all aspects of domiture. We can break the planning process into various stages, each with accompanying tasks and subplans. This plan, with all its partial plans, is necessary to protect the scale of landscapes that are too large to see, except perhaps by satellite, the parts that are too small to see, such as fungi and viruses, the parts that are too-long-lived for us to observe, such as long successional changes or evolutions, and the parts that we are ignorant about. Without special effort, we are aware only of what we see working in the system during a very short time. We trust that our plans will ensure that the system will remain as a healthy entity for a very long time so that many generations of us can gather our needs from it.

Planning is not meant to be a finished work of art—it has to reflect our understanding and use of nature. Each activity needs to be fed back into the process of updating the plan. Implementing the plan should result in improvements to it.

Protect Hotspots: Madagascar Hawaii & Ecuador. The Critical Ecosystem Partnership Fund (CEPF) is a major endeavor to preserve Earth's most critically endangered and biologically richest regions. The biodiversity hotspots are in a state of emergency, according to Jorgen Thomsen, CEPF's Executive Director and Senior Vice President at Conservation International, the managing partner of the fund, who stated, "By engaging local people in biodiversity conservation, we ensure the best chance of success at protecting the environment for future generations." The CEPF, a joint initiative of Conservation International, the Global Environment Facility, The John D. and Catherine T. MacArthur Foundation, the World Bank, and the Japanese government, aims to invest at least $150 million over five years in biodiversity hotspots—highly threatened regions where more than 60 percent of terrestrial species diversity is found on only 1.4 percent of the Earth's surface. The Commission would work to refine and coordinate protection of all identified hotspots. The areas identified here and below are representative samples, not an exhaustive list.

Keep Critical Areas Intact: Amazon Congo & New Guinea. Many areas are critical because of their size, as well as their uniqueness, and due to their out of scale effects on the global system. These areas should be kept in tact, as functioning systems, although they could be used by archaic cultures and by constrained industrial systems, if precautions were taken. Their use by nations would be limited. Earth parks in the Antarctic, Amazon, Arctic, Northern Canada, Congo, New Guinea, Oceanic areas, and Eastern Russia would be declared immediately.

Restore Large Areas: Mesopotamia & China. Ecosystem restoration would be begun; massive planting efforts are undertaken. No further expansions would be permitted for development in wetlands or other sensitive areas. Destructive searches for resources would be suspended, in favor of substitution and recycling. No new building would be encouraged until uninhabited ones are restored.

Many special areas, now degraded or destroyed, could be restored. Candidates for restoration would include: the Zagros mountain area between Iraq and Iran, the Shat-el-arab river system in Iraq, the forests of Lebanon, the central tall-grass prairie in the U.S. for buffalo as a Wildlands Project, and Northeastern China forests.

Calculate an Optimum Global Human Population. An optimum population for the planet could be calculated, using a deductive, synthetic, conceptual model based on data generated from research on net primary productivity and net community productivity, and on technological expansions, and on the limits of human cultures. When NCP is used to calculate a maximum population, by adding the calculations for each vegetational unit of the earth, the measurements are consistent with previous low figures—1.08 billion—after the same percentages of inedibility and waste are subtracted, and technological and cultural increases or decreases are added. The population of 1.08 billion is almost identical to a flat 1 percent rate of the NPP, subjected to the same loss percentages (900 million). The use of increased animal protein would reduce the number correspondingly. This availability factor is on the order of 75 percent.

The figure, 1.08 billion, is a maximum. An optimum population, arrived at by a 50 percent rule, is 540 million. This would insure against problems due to fluctuations in productivity. Richard Watson arrived at a figure of 500 million, based on American levels of consumption with the present industrial production. Daniel Kozlovsky intuitively estimated 500 million also as an equilibrium population. Others, such as Arne Naess, have suggested similar low figures based on traditional livelihoods and fitness.

There are also minimum viable populations to be considered. For the human species, it is unlikely that the lower limits will be approached in the foreseeable future. There are several lower limits to keep in mind, however, such as ideomass and genetic minimum. Minima could include genetic 2000 people, and ideomass 600 million. Maxima might be for wilderness 1.4 billion and for social advantage 2 billion.

Before an optimum human population for the world can be meaningful, other questions must be addressed, for instance: How much land should be left in its native state? Enough to save one of each kind of ecosystem? At what level of luxury should humans live? What are the physical limits of resources? And finally, what is an optimum? — Laboratory studies with rats show that, with a choice of optimum rat environments, some rats will reject the optimum. Many humans might, also. Is it meaningful to speak of an optimum for all of humanity?

How these questions, and others not asked, are answered determines an optimum human population. In calculating an optimum population within ecosystem restraints, we have to consider minimum wilderness preservation, air and water quality, genetic minima, nonrenewable resources, appropriate technological innovation, the importance of cultural frameworks, adventure, research, beauty, uniqueness, and other intangible experiences — although these dimensions cannot be quantified. The number 540 million people tries to address all these concerns.

Anticipate Climate Change. What is the solution for climate change? Restore forests and grasslands, and use alternative energy. Should we try a high-tech solution? A massive program, like the atom bomb, only for alternative energy technologies, might work. Continued global warming could lead to reversal of ocean currents. One way to anticipate change for preservation is in the design of protected areas, using a north-south axis and including many different elevations. Siting cities, or most of a city, over 20-40 feet above sea-level might be prudent.

Commission on Cultures & Religions
This commission addresses how to keep cultures healthy and active by understanding the characteristics of a healthy culture and meeting its needs, from being grounded to being sophisticated. Cultures change over time;

some develop, some collapse. Many cultures are transformed when they adapt to changes. Some cultures were transformed by domestication or agriculture.

Culture operates like nature, with rhythms of dissolution and reformation. Often the elements of a culture will simply be rearranged by a succeeding culture. A new culture can only be made from the heritage of the old. Our survival depends on the capacity to remake the image of the world from within, phoenix-like.

Like biological species, cultures do not fit perfectly into an integrated whole; there are discontinuities and contradictions. The culture is a loose-fitting patchwork of ideas, things and relationships. Humans can tolerate inconsistency and operate with contradictory beliefs: Soldiers fight for peace; ministers save the unborn for starvation. If the contradictions become too great and maladaptive, then the culture can collapse or disappear.

In the face of a change a culture can either embrace change or resist. Resistance to change is normal as a cultural process. Groups like the pygmies have specialized to fit the requirements of the environment, successfully. This makes it difficult to adopt other cultural arrangements.

On the other hand, resistance to change itself is an adaptive mechanism. According to Betty Meggers, it works as a successful "cultural isolating mechanism." Isolation remember is what allows a culture to develop in the first place. But, then does it force a culture to become stagnant? This Commission will examine the history and features of cultures and religions.

Preserve Cultures & Languages. The spread of the Bantu people in Africa caused a destruction and then a homogenization of languages. But then, because the societies were still relatively small-scale, they soon started to fragment in local mosaic environments. Bantu, for example, spread south as far as it could and people adopted that language, but then dialects started to diversify. Bantu now has produced 500 daughter languages in the past 200 years. Things tended again to a linguistic equilibrium. Of course, other peoples, such as the Hadza, Sandawe, or San Bushmen were pushed to the margins.

Language reflects places. Knowledge of the environment is coded in a language. Local peoples have knowledge from thousands of years of successes and failures. In Palau, they know the 1000 approximately fish species. The Kapingamarangi islanders in Micronesia spread their catch over 200 species without threatening their numbers. Western science has not had time to identify the species, much less create an effective marine management system. The vast undocumented traditional environmental knowledge is kept in indigenous languages by those people.

Languages can disappear for many reasons, normally, because people stop speaking them, especially if the young abandon the tongue and the elderly die. Occasionally, it can happen that a drought, famine, disease, war, flood, earthquake, tsunami, volcano eruption, acts of genocide, or other catastrophe wipes out a people. For example, the Paulohi language speakers in Maluku, Indonesia, experienced a severe earthquake and tsunami several years ago which killed all but about 50 of them.

Some people have said that they do not want to bring children into a world where their society, language, and people have no place. Some have turned to negative behavior like alcoholism, drugs, crime, or killing. The Waorani in Ecuador, the Carabayo in Colombia, and other groups in South America turned to killing, and for that reason some groups have still not had peaceful contact with the outside world.

This commission would work with groups that have diminishing number of speakers. It could help publish dictionaries, audio tapes, and news programs. It would promote groups that are successfully trying to recover their languages, such as Hebrew and Hawaiian.

Allow Religions to Act & Flourish. Religion is a part of culture that binds people to their ancestors as well as to the invisible powers of place. It focuses on the changeless aspects of natural and human processes. Religion concerns itself with an image of the world, that explains what the world is like. It also explains how we can influence it and why we would want to influence it. A shared religion affirms family and ancestral ties, but it also allows strangers without kinship ties to act more peacefully. Of course religion has also served more mundane human purposes, such as to justify the transfer of wealth to a leader or to the rich, or the sacrifice of lives for an ideal nongenetic reason. Religions, and myths, according to Joseph Campbell, are great poems, pointing through things to the ubiquity of a presence in each. Cultural inheritance seems to work. The heritability is high. Children tend to adopt their parent's religion, political views, and leisure interests.

Religions are attempts to understand or control the world, either by understanding the invisible or by having spiritual beings intercede. Religion tends to reinforce the integrity and structure of society, by providing a common image, and reinforce the belonging and commitment to the group. Religious claims about the other spiritual world tend to be counterfactual, but they cannot be too implausible. The supernatural has to play a part in the world. It has to be associated with living beings.

Religious rituals can also stimulate endorphins to the brain. These rituals may include painful poses, rhythmic movements, singing, or trials of endurance. Endorphins have good effects on the immune system. Trance states are another feature of religion. Endorphins allow people to feel positive about people who share their experiences.

Each religion is an attempt at transcendence with its own truths, certainties and stories. Their differences allow common beliefs, such as the golden rule, but also inhumanity. Some social virtues, such as trust, truth, restraint and obligation are grounded in religious beliefs. A contractual economy needs these virtues to survive, but at the same time, it is undermining religion with secularization.

The stories of religions concern events that are deeply meaningful to the listeners. This helps bind the group, also. Religion may help control disruptive forces, especially things about distribution and power. Religion coerces people into a social contract. Religion and story-telling may reduce variability of individuals in a group. But, this might increase variability between religious groups. Shared beliefs in a religious community may allow it to outcompete a strictly secular ones. It permits more sacrifice and commitment. Religion also allows ecological balance for many groups.

This commission would work to preserve the cultural capital built up by religions of thousands of years and regenerate other capital.

Create an Office of Ombudsman. Although not traditionally a fourth branch of government, as first proposed in Sweden, the function of the Ombudsman would be to improve the effectiveness and fairness of the global government while protecting basic human and ambihuman rights. Through the Ombudsman, nations or cultures could present cases of dispute for resolution. One strength of the office would be its power to initiate legal proceedings in behalf of any human and ambihuman constituents or nations. It would be an independent branch with separate responsibilities: To investigate complaints against the GU or its officials for wrongdoing; to investigate complaints against nations; and, to investigate GU operations.

Avoiding Collapse
Our lust for growth and conversion of the planet is rapidly changing the social and environmental orders that represent the natural capital of social and environmental evolution. This reordering is also constructing an accidental trap that leads to massive catastrophes that could destroy that capital as well as the basis for its renewal. We are already experiencing local catastrophes, such as earthquakes, fires and pollution, and a few regional catastrophes, such as tsunamis. And, we are aware that global catastrophes have happened in the past and are possible now.

We know that there are physical and ecological limits to the cycles and renewal of the planet. We know that we may have exceeded some of them. In complex, self-regulating systems very small changes have large consequences. In some cases, where conditions like drought are cyclic, in the Sahel region of Africa, humans expand during the good times, only to perish when the drought returns. In other cases, human activities, such as deforestation or overgrazing of herds, can cause weather changes. The large scale and slow rate of changes allows people to view the situation as natural, but once these catastrophes pass a threshold, the people and their cultures have been trapped by their demands, and only severe reduction or collapse can allow the system to regenerate.

We also know that not all catastrophes are short-term, single, small, fast, and visible. And, we are already in the midst of slow, long-term, large-scale, multi-pronged, invisible catastrophes. The climate is only the first that is becoming visible. Extinctions are another. Catastrophes can come in combinations, for instance, an asteroid strike, followed by volcanic eruptions and deep ocean releases of clathrates; possibly climate shifts, collapse of freshwater systems and droughts. Civilizations have collapsed.

If civilization collapses, the struggle back to a technological society will have greater limitations. Accessible minerals will have been scattered; the gene pool will have been greatly reduced. Then it may be too late. We have to act now. This is an emergency, requiring large scale, multiple approaches, with new technologies, massive conservation efforts, and microenergy solutions (which require participation), but not using old, unconscious assumptions and design traditions.

Fortunately our species is very adaptable. We have and can act on a global scale when we recognize the need to, but in this case the actions have to be coordinated on a global scale. Any one strategy, such as reducing the human populations — without making changes in distribution, flexibility, ecosystem health, equity, and many other factors — might not let us avoid collapse, but merely postpone it. Reducing the population would relieve stress on ecosystems and reduce the destruction of species and systems. But, if we continued growing economically and connecting tightly globally, then the catastrophe of a single key resource, such as cheap oil, could set off cascading pulses of contraction and collapse. We can change our strategies by reducing our size and impact, and by decomplexifying our societies and governments. We need a strong international organization to regulate the vast commons of the planet and to protect cultures and nations in a framework of equals. We have to accept that our labors will be more intensive and that our luxuries will be smaller. We can balance our global push to health with self-reliance and self-control. With international cooperation, the risks can be shared and vulnerable cultures can be helped with minimal disruptions.

Making Enormous, Simultaneous Changes
The enormity of military budgets suggests that we are willing to dedicate massive resources for one thing we deem important, although that is not sustainable. However, the military and its budget could be assigned to the domestic emergencies at home, for making transitions in energy and agriculture. We need mobilization as fast as what happened in WWII. When the auto industry was drafted to make planes and ships. When strategic goods, from sugar to gasoline, were rationed. That emergency restructuring was done in months. Why not try for 6 months for global mobilization, now? Restructuring could be done profitably, to increase global security.

The cost could be quite reasonable. Lester Brown and others have calculated the costs of social goals and of restructuring the earth. The social goals came to $75 billion per year and the restoration goals at $185 billion per year (compared to a world military budget of $1.522 trillion). If we added other social goals, such as equalization and reparations, infrastructure repair, or the deconstruction of some cities and industrial area, it might only double it, to $150 billion. Increasing environmental goals, such as the restoration of farmlands or wilderness might double that estimate to 370 billion. If we added the social costs of transition, including monies for UN to take over police actions, education and health, that might raise social goals to $220 billion. Setting aside more land for hotspots and common areas might reduce income $100 billion per year. The grand total is $690 billion per year.

Compared to $1.522 trillion per year, $690 billion is a half-price bargain. The military budgets are not affordable anyway, which is why social and environmental goals, and the infrastructure of civilization, have been so neglected. All that is really required is convincing world leaders that we are in an emergency situation, now, and need to act before a major catastrophe claims millions of lives and billions of property. Of course, they would have to sacrifice some money and some comfort, but then they could be heroes. And, that is what we need, uniform sacrifices for the benefit of all.

The expenditures, as well as coordination would have to be handled at the international level, by the United Nations, which would have to monitor progress towards the goals. It would need new offices and agencies to pay out the money allocated for it. The UN could also appoint new Boards to oversea each individual goal. This scale would be larger than anything done before, so it is important to have a hierarchy at the top, but let the smallest groups be loose and inventive. Sometimes the most effective directions come in a bottom-up way.

This scale of change, with actions performed at one time in every community might be possible. Everyone has to understand that it is a planet-wide emergency, as immediate as any hurricane or earthquake; it is just more subtle. There will be social and cultural hurdles. Of course, the rich and powerful now will not want to sacrifice the comfort of the status quo, but they will still benefit more than anyone else. It is in their interest to still be the richest group in a healthy culture and planet. If we can just educate them to their real interests. If. Some cultures will not want to work with other cultures, if they perceive unequal gains. That is why a diverse body of managers is necessary. No matter what the outcome, this has to be tried, for the whole planet not just for self-assigned, special groups.

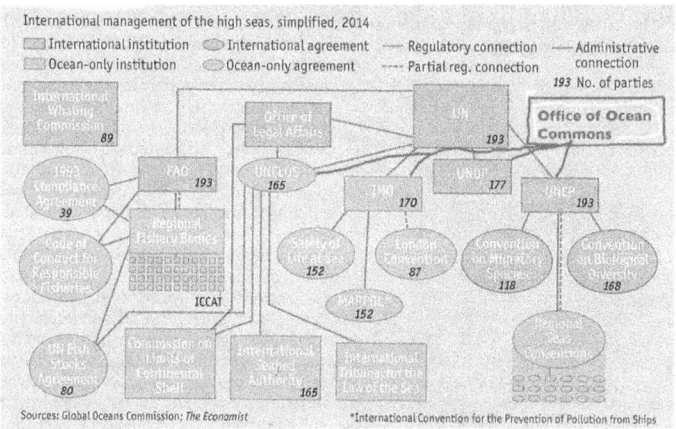

Figure 5. Office of Ocean Commons in UN (or GU)

The Future: Will It Happen?
 Yes.
 If.
But, 'if' always has so many complications. Taking these steps would solve many of the problems addressed earlier. The satisfaction of physical and cultural needs, as a result of living in stable and small societies, would contribute to the health of people. Fitting economic costs and needs to the limits of ecosystems and monitoring the economic process would reduce wastes and pressures on natural processes. The coupling of agricultural productivity to a solar budget, and the conscious restoration of degraded systems, would contribute to the health of ecosystems. Sufficient wilderness would allow the self-maintenance of global cycles. With the increase in security, wealth, and self-esteem, human populations could be dependent on ecosystem productivities and still be diverse and unique.

With the removal of war capabilities and the equalization of wealth, the remaining issues are not the kind to incite violent passions. Disagreements over the best way to raise wheat or maintain a forest may be more easily resolved than deciding the best nation or truest religion. The death of large-scale dogmatic ideology and national idolatry could also mean the end of organized slaughter. We have to perfect the art of resolving conflict. Mastering it through social debate would free unprecedented resources to satisfy social needs. Perhaps a planetary electronic referendum would open communication. In designing the world, everyone can participate. We can reduce the violence to nature and ourselves and transmute it to debate. That which has been hitherto left unsaid-what we want to become, what we could become-could become explicit.

Changes would be made after a period of adjustment. The process must be sustainable and equitable. The population would be adjusted for carrying capacity.

The real answer is still 'if.' If things get worse, if there are catastrophes and collapses, and if people try to take charge of the global course, then a Eutopian framework may be instrumental in creating a rich, equitable and peaceful future. But, if things really deteriorate, then a eutopian course is not likely to be attempted.

Figure 6. Highway & Shipping lane plans

Why It Could Work: A Brief Conclusion

The Dance of Art Money & Ethics: Advertising Good Places
Advertising creates the mythic images of our industrial cosmology—Marshall McLuhan called advertising the "cave art of the twentieth century." The myths are powerful, but trivial, and memorable, but inadequate to convey the meaning people need to live. Perhaps the myths are restricted by their content. If so, then ecologists and artists, as well as urban planners, historians, and politicians need to use the strengths of advertising to convey ecological sense and traditional wisdom, the feelings of balance and the dreams of nature.

Our dream of nature, and it is still a dream, in modern Western culture, is the dream of order and beauty. But, as Aldous Huxley noted, the dream of order begets growth and tyranny, the dream of beauty, monsters and violence. Our dreams are nightmares because they are not complete. The nightmares are symptoms that reflect unbalanced and immature cosmologies, that is, images of the earth. A traditional cosmology evolves with people's needs, fears, and knowledge. But, if it is incomplete, or if it does not fit environmental conditions, it may fail. Many early cosmologies, primitive or advanced, failed to fit the earth.

The modern industrial image of nature as a resource has resulted in pollution, material shortages, and environmental degradation. A culture that degrades its ecosystem risks its own extinction. Industrial cultures, however, are not the only cultures in existence. There are hundreds others, although at one time, around 1900, our species had over 1,000 different cultures and 3,000 languages, roughly equivalent to the number of natural biogeographical provinces and subprovinces on earth. Each culture exists in a particular location with a unique history. Later developments are not more adaptive than earlier; nor do they replace them. Ethnic groups are not evolutionary stages culminating in the U.S., but are equally valid ways of life. Each culture is only one of many possibilities. There is no correct way.

Each culture has a root metaphor. In the West, it is the machine. The advent of the machine made processes of order more amenable to description. Although only a closed system itself, the machine was a fruitful metaphor for living systems. The theory of the living organism as a mechanical contrivance explained biological phenomena from the physiology of an organism to the processes of cells. The cybernetic machine metaphor was successful at explaining detailed processes without answering fundamental questions of meaning.

Science makes extended use of the metaphorical process to construct its models. For example: "Man is a system" according to Ervin Laszlo or "Man is a computer" according to Michael Arbib. Kenneth Boulding offered the perfect machine metaphor for the operation of the earth: as a spaceship. As a metaphor, a spaceship suggests the limits of earth and the value of a limited life-support system—unfortunately, it also implies something of human creation that can be controlled and fixed by human intention.

The use of the word 'ecology' by Ernst Haeckel implied that the natural world was a place to live, a house, rather than a machine to control. Making the earth into a house is fundamentally a poetic activity, according to Gaston Bachelard. Poetry also is a way of understanding the universe through metaphor, a literary device that transfers the characteristics of one term to another. As Picasso said of art, poetry also is a `lie that tells the truth.' For example, Shakespeare said "The body is a garden" and Harvey said "the body is a machine." The body is not a garden or a machine, but the metaphors extend our understanding of the body.

Poetry is communicative of the quality of things. Like science, it discriminates the unsuspected in the commonplace. It is not different from science, but more diffuse; not better than science, but more comprehensive. It accepts ontological parity, the equality of beings; aspects of the world are not negated or reduced by one another. As metaphorical knowledge, which may be prerational or metarational, poetry can avail itself still of scientific references. Poetry can measure a whole qualitatively and mimetically, a germ or the cosmos with its imagery. Poetry is a tool for comprehending partially what cannot be known totally. A poetic language could include a view of the interrelatedness of all existence in a sublime ecology.

People need to be made aware of the power of self-determination. People need to feel things, like the immensity and uniqueness of nature or the strangeness of a biting tick, before they can act. Poetry can help people feel themselves as part of the web of life or on an oasis in space. That feeling, more than laws or injunctions, can justify preserving the ecological systems of the earth on which we live. Humanity is a poetic species, as Richard Rorty noted, "one which can change its behavior by the words it uses." We need desperately to change our behavior.

Mythology can join science with feeling to help us change. Mythology is not limited by method. Mythic symbols store information concisely, which makes it possible for a person to assimilate the collective experiences of a culture. Myth combines us with other beings. Mythologies are in fact great poems that function to awaken the experience of awe and humility before mystery, create a cosmology, validate and maintain an established order, and bring the individual into harmony with the whole.

Monetary Lies & Pecuniary Pseudotruth

Unfortunately, the myths of the predominate industrial cosmology are inadequate. The myths are powerful, but trivial and misdirected. Poetry and art are undervalued as forms of communication, not to mention as ways of shaping and making. Business has transformed much of art and poetry into advertising, to match the style and attention span of the people in industrial cultures. Advertising, quite literally from the *Wall Street Journal* to college textbooks, refers to its activities as "shaping the American dream." Like art, advertising creates an image of a way of experiencing. Unlike art, it limits its focus for a specific goal—profit. Like art, it mirrors us. Unlike art, it intensifies and glorifies only the positive aspects of culture, ignoring the dark, negative aspects that are equally valid.

Its simplicity is irresistible. Our environment deteriorates according to ecologists, but gets better according to economists. Their pictures are prettier. People want to hear that it is getting better. Advertising tells them it is. People want to act stupid and spend the inheritance of their children on themselves. Advertising tells them their actions are rewarded. The real issues of life and death, destruction and hope, make people feel helpless and anxious, so advertising draws their consciousness to comfortable trivia.

Despite the ugliness of the dreams of progress and growth, of waste and stylistic frenzy, advertising, using sophisticated techniques and narrowing the focus out of context, makes the dreams desirable and irresistible. People in agricultural and hunting cultures interiorize the abstract industrial vision. African farmers are convinced to buy inorganic fertilizers, even though it degrades the soil; women to buy powdered milk for their children, even if it kills them. Tractors replace draft animals in the paddies in the Philippines, even though they are costly and less efficient; French winter fashions are found desirable in tropical Brazil, even if they can only be worn in air-conditioned villas. People in industrial societies are convinced that their children will be ruined without personal computers. Disposability is offered as a fix to a void in the temperament. Advertising fuels the acceleration of conspicuous and compulsive consumption.

Ecological Persuasion
Yet, advertising may be the most effective means to reshape desires and reform buying habits. Advertising presents the symbols of modern experience, even if they are just the trivial ones. It could present healthy symbols equally well. Advertising does incorporate traditional values, like family, friendship, and love, although to sell beer and cereal and, sometimes, churches and hospitals. And, like art, advertising lies, although Jules Henry thought it was instead a new kind of truth—``pecuniary pseudo-truth''—not intended to be believed, or certainly, proved.

Advertising is beginning to support more informational functions, such as the dangers of drug abuse and smoking. Advertising creates values—fur coats, fast cars, dark beer, slim cigarettes are certainly recent and artificial values—but it could be used to create positive ecological values and new identities that show that our needs for prestige, esteem, and belonging can be met without stylistic waste at mindless speeds. Advertising could promote new attitudes about appropriate technology, the rights of other cultures, and the place of people in nature. Good advertising could be as subversive and conservative as ecology. It could avoid confrontation with people's values; emphasize positive aspects without negative ones. A good ad could capture and carry the most self-indulgent viewer; for the most part, ads don't require effort, just attention.

Advertising has been serving the dream of progress, but progress is leading to catastrophe, a long, slow, global catastrophe. When people experience local, sudden catastrophe, they usually respond immediately, with heroism and sacrifice, aiding the victims of earthquakes or floods, sometimes famine. Advertising could bring to consciousness the slow catastrophes of erosion and population overshoot, and, perhaps, invoke the same altruistic and effective responses to them.

To work towards this service, conservation groups could define and promote an integrative mythology as the basis for the framework of diverse efforts to protect life and the environment. Conservation groups could provide a meaningful philosophical foundation, as well as coordination for other humane, social, and conservation programs. But, the approach must be egalitarian: Respect for life cannot neglect human life and suffering. The approach must be eutopian: A new cosmology cannot ignore adaptive cultural traditions that arose in place over centuries. Furthermore, in addition to formal education, they could provide re-education through the most effective means, such as advertising. Conservation groups could spend money advertising 'humaneness,' moderation, and the joy of living, instead of just consuming or winning. Ecological ads would be unique and compelling, simple and effective. They would advertise not a product, but a way; not for a profit, but for a dream; not for comfort, but as emergency actions to combat catastrophes.

The other day, tired of scribbling, I went to visit friends, who were watching a car race. It occurred to me that only with entertainment industries is there so much technical fizz and coordinated enthusiastic teamwork. Imagine all that energy and enthusiasm directed to appropriate technology or the proper use of forests. Imagine television coverage of forest work with the same amount of attention and detail. Why not a competition for the most beautiful or productive forest or teams working to restore devastated areas — broadcast by a major network as an important event. It also occurred to me that this remorseless entertainment is an anesthetic against the fear of emptiness or self-searching or death. Continuous entertainment is a kind of guarantee of health, riches, and long-life. Everything that is pleasurable, thought George Orwell, seems to be an attempt to destroy consciousness. Ecoforestry, for instance, cannot ever compete with entertainment if it raises troubling questions or difficult expectations. As long as the forest industry can guarantee many forests through the arithmetic of fantasy, ecoforesters will always seem to be complainers and false prophets — until it's too late, then they will be blamed for not avoiding the catastrophes.

Maybe the situation is not that bad. Maybe we can present images that rival the industry images. Maybe we too can speak the languages of euphemism that large corporations use to conduct their businesses of larceny and fraud. Positive images and pleasing language skills are everything these days; no one really looks for substance. The devotion to money, beauty and youth is the focus. I think one way to compete might be to present conservation biology or ecoforestry as a medical discipline, aimed at restoring rivers and forests to health — and advertise it that way. As with any medicine, the patient actually does most of the work to become healthy, although the doctor gets the credit and the payment. This could lead to more respect for the practitioners, but also to more responsibility and more rules. The first rule, which is basic, is identical to the first vow of the Hippocratic oath, "Do no harm."

Creating & Maintaining Eutopias Now

To avoid fanaticism and violence, Karl Popper has suggested that utopians should try to build an open, progressive, partially-planned society, instead of a finished, closed, completely-planned society. Indeed, this is how general systems theory would describe a working, successful society. Such a utopia would have to accept the imperfect nature of humanity and the changing ambiguity of nature. Utopias is the dream of reason—Eutopias is the dream of small traditions and cultures, reasonable or not. Where an imagined utopia offers revelations promising a desired future, Eutopias offers references from selves and cultures for producing good places on earth now. There is no mechanical prescription for making good places, no blueprint or timetable. The current institutions cannot create good places; the market has not been able to create health and equity; even radical ecologists have not been able to create a way—Eutopias is a fourth way. It is not an institution that benefits only the rich; nor is it a schedule of temporary handouts. It is a plan for a framework for local self-reliance and global exchange, that is respectful of traditional cultures and ecological networks. So far, there is only the idea or poetic image. Human will to power might be found in the will to imagine, and then to speak and become. What is our moral responsibility for this power? We can choose to alter our world with new images moderated by new ideals, such as good places, Eutopias. Eutopias should offer knowledge and power with charity.

The criteria for Eutopias include: Its benefit for humanity; the inadequacies of the present system; a drastic system change as a result of catastrophic awareness; and a low, but not too low, political feasibility. The benefits must be worthwhile to justify the costs. Benefits cannot be vague and unsatisfying when the costs are immediate and painful. Poetry and education must prove the benefits, so that the eutopian alternative can be begun. This code emphasizes its flexibility.

The best thing to do is to stop growing, stop producing, stop running. Suspend the race and contemplate a direction for a while. We know that whole countries have built again from ruins. So there is nothing to fear from stopping—if we know there would be no problem going again. What are the dangers of fast social transformation? Lack of justice? Lack of order? That is what the Eutopian framework is for—to avoid those lacks.

The steady state would provide a period of rest, and time to explore human values and quality considerations. Perhaps, as a dietary consideration, we could freeze energy use at 1890 or 1926 per capita levels; those were times of relative or optimal luxury for many people after all, and we have much more efficient appliances, now. This strategy would avoid eventual hardening of the choices. But it must be instituted at once. The crisis caused by exponential growth and destruction cannot be solved just after some final limit is passed and a great catastrophe has begun. The crisis of ignorance cannot be solved by hurrying and creating more problems.

Eutopias exists in the extended present, incorporating past traditions and future values. It would concentrate earthward (down) and inward.

Heaven may be a perfect home; Eutopias is here and now. Eutopias is a new comprehensive philosophy to make sense of the world. Eutopias is comprehensive and global. A broader frame of reference is assumed. It is concerned with unfolding and producing new emergent forms, not just a static description. Eutopias is grounded in environmental concerns. Its values must be highest cultural and natural values. It must develop from existing social and political forces. Eutopias is vitally concerned with the well being of society. It regards society as a *sui generis* entity, not just an aggregate for analysis. Conservation is a means to an end, which is human fulfillment in harmony with nature. Human happiness depends on a balance between needs and commodities. In a throughput process it is not possible to economize all inputs simultaneously. There are many criteria for inputs to be preserved. Options must be site and culture specific. Eutopias recognizes and preserves slow cultural knowledge. A social base may be partly developed through ecological education. Social diversity may have cross-cultural appeal. Eutopias would retain the capacity to change and innovate, with changes in environments and human values.

Eutopias would detoxify national rivalries. Racism, sexism and ageism would lose their importance in a cooperative society of advanced communication, automation, equality, humane scale, and meaningful preservation. Eutopias is politically aware. We make political statements by the way we live. Every tradition is only one tradition among many. The higher sanity requires of philosophies and therapies is open, planetary dialogue between modern experience and sacred tradition. Eutopias requires a planning process that bridges all cultures and sciences. It must be a participatory political movement. It must appeal to a large segment of the total human society. Since not all interests will be satisfied, there must be opportunities for transformations or alternate paths. Eutopias would be a framework for microeutopias, where different human experiments are tried. Its variability would insure that we could reject any local visions that fail.

The Eutopian frame can be justified. It is not just one kind of global society in one place. If we try to make one society, it will change over time, because people are in different places. Solutions come from living in place.

How can we form society within an ecological perspective? By following the principles of ecology and applying them to the characteristics of good places. According to David Orr, certain design principles work with ecosystems and nations. Small units dispersed in space, redundancy, short linkages between modules, simplicity, diversity of components, self-reliance, decentralized control, large margins, quick feedback. A megaframe like Eutopias would allow this.

Eutopias requires a change of attitude. We have to change the framework so that we can change to new minds. We no longer have an external point of view. We are inseparable from the environment and each other, but we differentiate. Eutopias is a nostrum, really, a description of good places for all beings. Its connotation is as a panacea or questionable remedy. This is appropriate since a panacea is a cure-all, a remedy for disease, and a solution to catastrophe.

What a Eutopian Approach Can Do

Eutopias is a self-conscious panacea. It requires an understanding of the anatomy, physiology, and diseases of the body in question, now the entire human and wild planet.

Eutopias Can Eliminate Bad Approaches & Actions
Human approaches to the challenges of nature have resulted in many losses. The lure of size and simplicity has resulted in many failures that become traps difficult to avoid or leave. The stresses from these things have resulted thefts, as attempts to balance or correct the situations. The very size and impact of humanity has made theft the only easy option, much easier in the short run than planning or self-restraint.

Eutopias Can Reduce Losses
Eutopias can reduce the losses of nature and culture by creating a framework to protect them. Eutopias can reduce the losses of health, fitness and accord, by emphasizing them and creating circumstances for their continuity. Eutopias can reduce the losses of equity, renewal and design by offering new designs that allow for a normalization of equity and for the normal processes of renewal.

Losses from accidents and diseases can be reduced by preparedness. Losses from earth and climate changes can be reduced, also, with preparedness for 'normal' events, such as hurricanes, earthquakes, and droughts. Design can also be used to reduce impacts from these events; for instance, by denying building permits on floodplains. The losses from some events, such as droughts resulting from El Nino, can be ameliorated, by having surplus food and supplies stockpiled.

Eutopias Can Reduce Thefts
Stopping a theft can be as simple as stopping a thief. But, theft has become such a complicated thing, many steps removed from the people who make the decisions and from those who carry them out. Eutopias would address the processes and trails of theft.

Reduce Theft of Life from Democides and Ecocides
Animal and plant lives are stolen for the profits of a few; ecological and human systems collapse as a result of biocide and ecocide. Hundreds of millions of human lives are stolen every century at the behest of a few. Millions die from starvation when regional crops fail; millions more die when the distribution system fails or is perverted for a few. This democide is unacceptable. The plague of power is responsible for the dialogues of death, and the absoluteness of some power is responsible for the massive scale of deaths in Russia, China, and some other nations.

Three reasons for these deaths seem more contributory than others: Inequity, runaway cultural antagonisms, and the use of absolute power. Inequity is simple to understand; some people have more valuables than others. Often, equalization does not happen until during or after a collapse, as it may have happened with the Mayans in 795 AD (or 1211 YBP). Cultural antagonisms seem to worsen when the different cultures are forcibly combined in nations as the result of colonial wars. Totalitarian regimes, especially with great power and in secrecy, behind walls or threats, kill almost as many people as famines — more, when one realizes that many famine deaths come from denial of distribution. Famine itself is the regional failure of food production or distribution. However, the denial of distribution, as when food is saved for trade or the elite, such as the English did to the people of Ireland and India in the 1800s, can contribute to the severity and extent of a famine.

Eutopian responses to these three reasons are: To equalize wealth as much as possible, to separate cultures into separate nations, and to make laws to control ecocides and democides. Laws without supervision and enforcement would not be effective. The UN outlawed war, but when the U.S., U.K., Iraq, Korea, or others decide to violate it, no one enforces it. The GU needs to have the power of enforcement, backed by all member nations. The GU needs to have the moral force to keep national governments open and responsible. The workings of government should be transparent to the people, whether the government is democratic, royal or charismatic. The GU must be sure that people can bear witness to every regime in the world, from Korea and Cambodia to the U.S. and Russia. Bearing witness should reduce the number of secret pogroms. Restricting and checking the power of leaders, in a Eutopian framework, overseen by the GU, should greatly reduce democides, including genocides and mass murders. There are no permanent solutions. Eutopias can reduce the theft of life by making it more valuable and more visible.

Reduce Theft of Common Sense
Eutopias can reduce the theft of common sense by fostering and respecting common sense in communities and nations, as well as increasing its value and appreciation.

Reduce Theft of Choice
Eutopias can reduce the theft of choice by creating a framework that offers more choices for people in different communities and nations. It can also address the economic and political processes that reduce choice.

Eutopias Can Reduce Failures
Eutopias can reduce failures through education and opportunity. Perception, intelligence, and imagination can be taught. Integrity, will and charity can be shown by example. And, if there are enough teachings and examples, these human capabilities will be developed and applied to domiture, that is, civilization and nature. With will and imagination, people can design and build good places.

Eutopias Can Integrate Tools & Designs
Tools and designs are important extensions of the human mind. Their purpose is to foster and assist survival, not to make it more difficult. Tools and designs can be made appropriate to environmental limits and cultural preferences, both of which are often ignored by industrial approaches.

Eutopias Use Tools Effectively
Eutopias can illustrate how people can use tools with awareness of their effects and impacts, and use tools with caution. The Precautionary Principle is followed.

Eutopias Can Thread Characteristics with Plans
Eutopias can thread the characteristics of nature, that is fields, ecosystems, and places, with the characteristics of cultures and good societies to make good places ecologically and culturally.

Eutopias Can Avoid the Traps of Noplaces
Eutopias can avoid the lure of or the accidental assembly of no-places. By exposing the lures of no-places, and showing the connections that make no-places into traps, Eutopias can neutralize the plague of placelessness.

Eutopias Can Suspend the Designing of Noplaces
By promoting the understanding of the inadequacies of bad characteristics and bad designs, Eutopias can stop the plague of uniformity and paucity. Through an understanding of the consequences of human ambitions and actions, Eutopias can avoid many of the evils that result from a civilization on technical autopilot.

Eutopias Can Increase Understanding
Understanding is a powerful thing. Understanding why an animal bites can dissipate the desire for revenge or punishment. Understanding why things break down can result in an examination of the context and effects of tools, and maybe a simplification.

Eutopias Can Help Understand Ways of Knowing
The ways that human beings know things is part of the human adventure. There are many ways of knowing, from traditional ecological knowledge to the most abstract science. None of these ways is the only way. None should supplant the others entirely. This is the importance of education, that it be applicable to local place, yet broad enough to put that place in a larger context, of the environment or planet or universe. These ways of knowing allow people to learn the operation of nature and fit human activities in it.

Eutopias Can Help Understand How Things Go Together
Ecology is a science of relationships and patterns. Understanding the components of an ecosystem, and the characteristics of an ecosystem, can be applied to the growth, change and development of human systems.

Eutopias Can Help Understand How Things Happened
History can allow understanding of the regular patterns of human life, as well as the dramatic changes, such as agriculture or urbanization, and how these changes have affected the patterns. History is a record of the cumulative human impacts on living systems, as well as of a few famous people or battles.

Eutopias Can Help Understand How Things Renew
Systems automatically renew themselves, especially living systems. Systems that do not renew themselves very well, such as agriculture and cities, can be put on track for renewal by linking them with the surrounding natural systems. Eutopias can foster the renewal of human systems by integrating them into natural self-renewing systems.

Eutopias Can Help Understand How Nature & Culture Work
Nature and Culture are systems. Domiture is the combination of those two systems. Culture was once called a 'Second Nature,' but human culture has expanded so dramatically that the two systems are better identified as one developed system, now. The fitness of human systems are intimately related to the fitness of species and natural ecosystems. The human attachment to place is critical to understanding why people live where they do.

Eutopias Can Help Understand How to Live in Place
Eutopias can offer understanding of how people live in places, not only how they adjust themselves to a place, but how they adjust the places to their needs and desires. This mutual adjustment can be ruinous or beneficial. Emotional investment in a place, even love for that place, is crucial to the preservation of the genius of place.

Eutopias Can Start Making Good Places
Eutopias can start making good places by addressing the economics and politics of human cultures. Economics and politics are large-scale human programs to relate human needs to resources and distributions of resources and goods.

Eutopias Can Show How to Preserve & Restore, Design & Plan
Eutopias can provide an ecological planning process that offers a structure of limits and divisions for the planetary system that would permit the preservation and restoration of natural cycles and places. An ecological design process would be applied to ecosystems as well as to cities and fields.

Eutopias Can Illustrate Ways of Making a Living
Eutopias, through a holistic examination of how people make their livings in place, can show how changes can make better places. Economies can be as diverse as tropical or desert places; there is no evolution to one economic style, such as capitalism. Eutopias can show how to integrate individuals,

communities and corporations into place.

The current, dominant economic and financial order is unfair. It needs to be radically reshaped. It is better to do this as part of a directed plan than after some kind of collapse. Power and wealth must be more equitably distributed. A Eutopian plan would try to ensure that the underrepresented would be allocated more resources. Which would increase the demand for basic services. But, would that let them disconnect from any dependence on a world market? Wouldn't that be the goal for every nation? To be self-reliant in terms of food and basic production? Then, nations could reject technologies and products that did not fit their cultures or that would affect their own resources.

Economies can emphasize different things, from producing family needs (reciprocity), to distribution and redistribution of luxuries, to trade (mercantilism), to unbounded capital and to bureaucratic efficiency. It may be time to emphasize a form of aesthetic efficiency, that is the shared production of what is wanted without as much regard for cheapness and mass production.

Eutopias Can Explore Ways of Governing

Eutopias can examine how people govern themselves. Political styles can be equally diverse; there is no evolutionary path to one political style, not democracy, socialism or community anarchy. Eutopias can suggest ways to fit governing to culture and place.

Politics can use different types of leadership or rules. The rule of law is fine. Perhaps a rule of religious tolerance would allow people to live together, something on the order of the golden rule. Does it always come down to one person, president, pope, king or dictator? Representing all? In small groups or communities, anarchy could work fine.

In a nation dependent on animals and plants and microorganisms in an ecosystem, it is important to give every being a voice in the changes that affect everyone. We are already used to representing the interests of the young and elderly, the handicapped and feeble; it is no great leap to recognize elephants, fungi and forests. This 'rule' of all beings could be called a Panocracy. This reapportionment of all the resources in a region would be enhanced by the drawing and making of ecological zones, which emphasizes the relative isolation of wild and artificial areas. This reapportionment of 'resources' that human communities have already claimed, as well as of resources that have been badly distributed as a result of theft or violence, may cause some degree of discomfort for wealthier people, but that is minimal compared to the suffering and death under the current system, which encourages overconsumption and immoral differences in the distribution of wealth.

Rather than rule, perhaps understanding. Perhaps knowledge. Would a rule of knowledge work? Or would people have to know too much? When it comes to politics, as with mythology and religion, different people have different levels of understanding. So, government should be simple enough for everyone to participate knowledgeably in.

Eutopias Can Try Ways of Integrating Religion & Art
Religion is a part of culture that binds people to their ancestors as well as the invisible powers of place. It focuses on the changeless aspects of natural and human processes. Religion concerns itself with an image of the world, that explains what the world is like. It also explains how we can influence it and why we would want to influence it. Art is a part of culture that expresses the invisible parts of society and the environment.

Religion can lead to understanding of the world; it can lead to ecological balance. Art can lead to peaceful ways of interacting with nature and other human beings. Art ruthlessly examines society. Religion reinforces the integrity of society. Art is a survival technique for humans on a wild planet. Religion relates the human to the wild.

Eutopias Can Start Making a Framework for Revolution
One premise of Eutopias is that many things have to be changed simultaneously; some things have to be eliminated, and other things have to be invented. This revolution in thinking and acting, especially on a global and national level, will have to be governed by the consent of people in those nations. The structure of human life, in unique cultures in specific places, the basic everyday experience of human life, will basically remain the same, but the superstructure, that is concerned with global trading and distribution and taxes—that will change.

Create a Global Government
In a revitalized United Nations, or a new Global Union (GU) or International Association, a police force would be used for positive nonviolent interventions to help people with problems or disagreements. Such interventions would be cooperative efforts by neighboring nations, coordinated through a revitalized Association. All states now have armed forces, whose primary duty is killing their enemies, internal or external, in unending conflicts. The unarmed police force would have different goals: Rescue from catastrophes such as earthquakes; civic assistance, such as vote getting or monitoring; and simple police action, being a persuasive presence in areas of conflict. The GU would insure the inviolability of the police to go anywhere on assignment and to intervene in any conflict when asked by any party. If the GU had most large-scale weapons, it might short-circuit the vicious cycle of armament races, and it is conceivable that the GU would need to use weapons in some circumstances. The UN has successfully used police for the observation of peace and for enforcement, in Cyprus for instance. The UN has used police in response to natural catastrophes, such as earthquakes, in Peru and Italy for instance.

A global association would also coordinate the distribution and use of common resources, which would also be owned by the association as representative of all nations, rather than of individual nations; resources across the planet are uneven and have precipitated numerous disputes for thousands of years. The agency would address real global problems, such as global warming, which has resulted in grain harvest shortfalls in recent years. Climate itself would be a concern.

For any nation, the association could advise on topics ranging from justice to wilderness. Wilderness has an important role in human freedom, as well as in providing ecosystem services. The association could insist on self-reliance, by connecting human population to ecosystem productivity. It could also make sure that local air, soil, and water resources are stabilized.

For all nations, the association could provide education on health and appropriate technology. It could work to provide basic needs for food and health. It could insist on the truth of the ecological situation, on the real costs of economic decisions and growth, especially those that destroy ecological capital, in the form of wilderness ecosystems. It could recommend changing the system to allow taxes on destructive activities, such as excessive carbon emissions, and to normalize the values of resources and wilderness.

Encouraging & Fitting New Nations
The GU will offer any culture the opportunity to have a vote in the global management of human cultures and natural processes. Those cultures may choose to remain in their current national framework and share one vote, or become independent and exercise a whole vote.

Three levels of responsibility — individual, national, and global — are identified and discussed; each has responsibilities for specific attributes, such as population or health. This does not mean that only nations will exist within a global framework; alliances and networks will form and reform.

The Eutopian code divides the earth into zones for preservation, conservation, domestication, and human communities. Human activities are limited to specific zones and, within those, the global authority controls all air, water and land use under complete sovereignty. Political units are formed from existing cultural units; an optimum human population is of each nation is based on a calculation of net community productivity on arable land through traditional agriculture. Common planetary resources are assigned according to the optimum population figure. The development of the nations is regulated by the Global Union through charters. Self-reliant nations decide their own appropriate technology, crops and institutions based on traditional values and are responsible for the ecological education of their constituents. Residents of nations have equal rights and work opportunities, and have the responsibility to participate in government and to live as wisely as possible, to make good places.

Address Connection & Size
When small societies start to grow, they become successful in different ways. But, that success leads to increases in size, which can lead to a tragedy of scale. Each major technological breakthrough permitted a step increase in size. The size of a local population increased the likelihood of its success. For cultures, size was important. More successful cultures (as measured by size and continuity) were larger and more aggressive. Humans naturally increase the size of societies, but do not know how to stop or limit it. Henry

Simons designated the great powers as "monsters of nationalism and mercantilism" and suggested that they are the obstacles to world peace, and must be dismantled for us to survive.

The processes of social development can create traps that then determine the direction of further development. Changes in scale, such as population size, can decrease the number of options possible, while providing different attractive options, such as the accumulation of luxuries. In this sense a trap is a sudden reduction in options or flexibility due to changes in scale or repetitive pattern.

The size of a culture might not be optimal, since they grow unplanned and wild. Kohr quotes Arnold Toynbee as linking the rise of universal states to the downfall of civilization. Toynbee suggests that one solution might be the return to the Greek ideal of a self-regulating balance of small city-states rather than further macropolitical solutions. Leopold Kohr sees gigantomania as the economic problem of systems. Kohr notes that our choice is not between crime and virtue, but between big crime and small crime, not between war and peace, but between great wars and small wars.

Kohr's reasons for the advantage of small states: There is a cultural diversion of aggressive energies — or artists are cheaper than soldiers; there is a relief from social servitude, as a result of time and leisure; there is the variety of human experience; and there is the testimony of history. Kohr concludes that it was always[17] "the small state, not the empire that survived. That is why small states do not have to be created artificially. They need only be freed."

Local equalities are easier to acquire first, more than global equalities. We live local lives in home places, with local limits and local pressures. We can calculate an optimum size population of a culture or community, by relating it to the carrying capacity of the land and desired properties of society. We can establish optimum scale of populations through limits of carrying capacity. We may also calculate an optimum size for the planet, based on the sum of local cultures. An optimum global population might be of the order: Between 0.5 billion and 1.5 billion people, the sum of local optima (Wittbecker 1983).

We may not know what is the minimum, optimum or maximum use of an ecosystem. Science might try to identify minima or maxima, but management can aim at optima or satisficia — Francisco Varela analyzes the evolutionary process as satisficing rather than optimizing; a suboptimal solution is adequate. A free market has to be limited by conservative calculations of ecological balance. It is almost impossible to estimate the economic value of natural balance.

Tune Connection & Speed
The speed of our economies might not be optimal, either. As new intense living in cities and new inventions for communications accelerate, our interactions and requirements for resources also accelerate, forcing faster economies and pressuring slower, traditional economies to try to compete in speed or risk becoming uncoupled from the fast lane. Yet, it might be

good for countries to be uncoupled. Uncoupling economically might be a sound option for traditional societies unwilling to make the same mistakes as industrial ones. Local communities are based on traditional cultures, which have long-term lasting power. Traditional cultures often have wealth-leveling properties, absolute property ceilings, fixed wants, and production coupled with need — all of which results in a stable economy. Then, also, efficiency and productivity are less important than use and appropriateness.

In the past, traditional communities have had a rich biological knowledge of animals and plants that allowed them to find or grow edible and medicinal plants, and to make appropriate houses and cooking utensils. This ecological knowledge is being relaced in modern societies by a rich economic knowledge of virtual animals and plants in computers and games, as beneficial insects are poisoned and soil erodes.

The satisfactions from being in a culture in place, from planting trees, growing apples, watching birds, playing with children, and making love are primary. And, they are not speed-dependent. Some things have proper speeds. Music is not made more efficient or better by increasing the revolutions of a disk. Food or relaxation require a human speed. We sophisticated moderns lack the wisdom to act as if we believed this. Those who are uncomfortable with primary meanings tend to become addicted to power, speed, and possession, as a frantic way to avoid awareness, silence, or responsibility, to avoid being grounded in nature. As modern cultures become accomplished in the secondary meanings of life, from money and economic success to accumulating goods and moving fast, the primary meanings of living in place tend to be discounted.

Dismissing nature in disgust, we attempt transcendence through speed. We speed away from nature, from our own bodies, and base our civilization on that momentum, praying — *requiring* — that it never stops. People's souls die, but secure in their power, they manage the things of civilization and inhabit the treeless flatscapes of the malls of commerce, comforted by the banishment of wilderness and the capture of animals in zoos and of free people in reservations, satisfied that their young are mercilessly tied to televisions and computers, acquiring information without touch and speed without grace.

Perhaps if we remain unconscious, there will be a shift to the fastest that will homogenize and level all human cultures. But, we can consciously imagine alternatives and work to preserve cultural and natural diversity and the richness of existence. Nations and communities do not all have to follow the same path and the same rules at the same time and at the same rate. Cultural success is not the 'survival of the fastest' any more than it is of the biggest or newest.

We have the knowledge to save cultures, to restore places, to participate in the cycles of the earth, but extra speed and power are not required. The pace of nature is generally balanced and well-established; we violate it at our risk. If we adjust to the pace of the growth of trees and to the movements of animals, we would not be risking catastrophic extinctions and famines, shortages of water and fuel wood, and the death of humaneness.

We do not need to give up power to faster, larger economies. We need to shift power to local communities through self-reliance and participation. A community protects individual freedoms, guards regional culture (values and identity), and holds groups accountable for their use of power. In communities, people can decide to be conservatively sustainable or to grow and gamble on innovation. Communities can have different economic attitudes, paces, and goals. A community that is balanced and flexible, in tune with natural cycles, based on traditional values — in which industrial production is limited to appropriate goods — can absorb the shocks of change far better than an immensely big, powerful, accelerating, postindustrial, national, global-tied vehicle.

Providing Paths for Individuals & Communities
Isolation can dangerous, whether it is isolated theoretical knowledge or an isolated culture. A maximum isolation can be bad. A minimum can be bad. We can strive for optimal solutions and control, but should settle for suboptimal and partial control, a satisficing solution.

A Eutopian framework suggests small solutions to big problems. Protecting and restoring ecosystems is a local effort. Reducing gases that contribute to climatic instability, reducing consumption in general, reducing the human population, and reducing conflicts, which contribute to the escalation of wars, are local efforts. Integrating food into ecosystems, to regenerate soils and repair ecosystems, integrating technology into a culture, integrating economies into ecosystems, and equalizing wealth, are local efforts. Although not every global problem has a local solution, people will have to address them in small ways, too. Climate would be very difficult to change, much less control on a global level.

Promoting Health at All Levels
Positive health results from being on good terms with cosmos. The idea of a right to health should be replaced by moral obligation to preserve health. We need to become attuned to the earth, to commit our fate to nature, and not just say that we have faith in modern technology to save us with an artificial environment. We must be flexible, not detached or noncommittal. We must commit ourselves and be able to adjust to necessary changes — to be in a state of risk. The harmonious interplay between humans and environment results in adaptive fitness, which requires a constant expenditure of effort to maintain. We must maintain an environment for plants, animals and humans that is healthy for all; this is an ecological approach to health.

An ecological approach must create flexibility and then prevent civilization from immediately expanding into it. Flexibility is uncommitted potentiality for change. Flexibility must be distributed among the many variables of a system. Freedom and flexibility in regard to most variables is necessary during the process of learning and creating a new system by social change. There are still many possible futures for the earth and humanity, but they become fewer as we burn or destroy the earth's flexibility and our options. Recommendations to reserve flexibility must be tyrannical.

The ecological health of a civilization depends on a single system of environment combined with high culture in which the flexibility of the civilization must match that of the environment to create an ongoing complex system, open ended for the slow change of even basic characteristics. High culture is not a return to the innocence of the Inuit, or the sparseness of the Cro-Magnon cave. It includes necessary institutions for the arts and sciences limited in transactions with the environment. Flexibility is needed; within limits, a variable can move to achieve adaptation. Health is the capacity of the land and water for self-renewal. Conservation is the effort to preserve this capacity.

Although the nature of the biosphere is largely determined by evolution, by organisms adapted to specific parameters and to each other, the anthroposphere tends to be artificial and managed, with only human needs considered. We need to keep as much of the natural world as possible in the anthroposphere; there is a human need for variety, individuality, and challenge to understand the nonhuman. Immersion in trees and bees is necessary to nourish human attributes that are in short supply: Awe, compassion, reflectiveness, and brotherhood. As humans move from concrete to trees, there may be a profound transformation in a scientific return to animism. The metaindustrial culture is one in which the trees are counted in a census of members of a community. In the shamanistic tradition, people are not viewed as individuals, rather their history and experience is seen as result of being part of the group of living beings.

One significant book in the Hippocratic corpus was *Airs, Waters and Places*, which showed how well-being is influenced by the quality of air, food, land, and general habits. It is as important to know from whence your body atoms and molecules came, as it is to know the history of a used car. Atoms that came from stars and rocks make up molecules of seeds, flowers, defecation, and rotting leaves, which are cycled through our bodies. Bodies are open systems exchanging materials with the whole environment. It is therefore important to choose carefully what is put into the body. Good food comes from healthy plants and animals, unprocessed and unpoisoned. Cells and enzymes react poorly to poisons and preservatives. Physical, mental and spiritual well-being are dependent on a healthy diet. As much as possible, one should know the origins of one's food — the soil, the plants — and be able to determine what becomes you.

Integrating Actions with Poetic Wisdom
A planet that is mindless is not entitled to moral or ethical consideration. The earth has a mind but ecocrises are driving it to madness. The alternative to ecological insanity is wisdom. Wisdom is the functioning of a mind that is respectful of its own boundary and processes, according to Gregory Bateson. Evolution is trial and error process of learning; all learning contributes to evolution of global mind. So the cure for ecocrises is the education of minds.

We need reeducation in the demands and recompenses of a sane, realistic world. Peace of mind, security and self-respect must arise out of being someone in a real community, and these will be valued more than conspicuous possessions and idleness. The process would be labor-intensive rather than capital intensive, and intimate rather than pretentious.

A high culture must respect the wisdom of its experience, using necessary technological devices (computers, televisions); be diverse enough to accommodate the genetic and experiential diversity of people; shall limit transactions with the environment, consuming natural resources (capital) only to make necessary changes. High cultures must depend on renewable resources from photosynthesis to wind, tide, and sun to continue 'making,' which is the source word for poetry, good places.

Poetry expresses the image of human potential, of what other circumstances may have formed. Poetry tells of a goal, even if it is the moral superiority of suffering in the 'third world.' By presenting a goal, poets can become the "unacknowledged legislators of mankind," as Shelley defined them. Poetry creates a fourth world, of groups sharing part of the wealth of the earth in a global community. This fourth world is where the past is reconciled with the present and the terrible beauty of the future is born. The terror of beauty, as Rilke recognized, results from its power to shake humans from the refuge of a small identity into an immense strange world.

The lives of humans and all beings has become a collective responsibility. Humanity has to learn to live again on a finite and varied earth. Learning is a transforming experience, but difficult. Poetry objectifies conscious experience and makes it easier to communicate. Romantic nationalism (in the 1930s) used poetry in service to the state. Used for each nation in this way, poetry can show the diversity of human experience. The essential unity of the earth can only be discovered through its infinite diversity. Poetry gives groups and individuals their identity; it articulates societies and authenticates forms of exchange. By transcending the limits of single cultures, it could draw all cultures together. The tradition of poetry does not belong to just three worlds; it encompasses them and links them together in a fourth. Poetry is wise language.

Figure 7. Proposed Antarctica Pyramid Arcology

Moving Forward Backward Inward & Outward

Human ills cannot be cured by a return to idyllic hunting and gathering groups or to a quasi-agricultural, ecologically-caring society. There is no possibility of complete return. Most industrial nations are urban, and are becoming more so, as agricultural countries pack their surplus peoples in cities. Nor can there be a return to 5th century B.C. Greece, or to 15th century China, or to 1910 France, or to any time. Many traditional cultures no longer exist; others are disintegrating under pressure from industrial cultures. Nor can there be a jump to a complete technological future, where technology transforms hydrogen into wealth for everyone. But, a Eutopian framework could move forward with traditional cultures and realistic ecological planning.

Uncertainty & Incomplete Knowledge
There is uncertainty in dealing with large, wild, complex, long-lived entities like ecosystems, or corporations, or nations. Leaders and managers have to live with uncertainty; this means that management decisions are essentially gambles. Gambling is a profession that acknowledges the operation of chance and makes conclusions in the absence of facts — few people are successful at it. This is an important admission, that we do not have facts to base our actions on, that nature is a stochastic process, and that ecosystems always changing. Furthermore, we do not know for sure what effects our actions will have on ecosystems, which used to live for so long, in such diversity, in so many places. Successful gambling suggests that the proper attitudes for gambling with nature are awareness, humility and courage, not arrogance, fear and maximum use.

Because of the uncertainty of human actions, ethics has to encompass the far past and distant future. No one knew that when DDT killed mosquitoes, it would concentrate in the food chain to kill birds. Values are time dependent, and ecological time can be very long indeed. The futures we invent are viable only if compatible with constraints imposed by the evolutionary past. An ethics that requires a long-range responsibility also requires a new humility, since technological power exceeds the ability to foresee its consequences. An ecological ethics recognizes the moral obligation to leave the world habitable for future generations.

The detailed planning of complex open systems is not necessary to create a good future. Planners are not in a position to attempt detailed models of future situations because many relevant parameters remain unidentified, and many of those known cannot be quantified. Plans can be made within the limits of variables, although it is not safe to be limited by lethal variables, as Gregory Bateson recognized. Closeness to real physical and ecological limits reduces flexibility, that is, the uncommitted potential for change. To minimize untested conclusions, Eutopias is based on the values and forms of traditional cultures. This could allow time for rational planning to catch up.

We have to invest and cultivate our inheritance. We must enlarge our human identity, to include other beings and the earth, to include our own posterity and its image of the future, without which we lose the will and capacity to solve problems. Creating the future is necessary to maintain the present. It is meaningful to construct a world that we will never live to see, to plant trees that take two hundred years to mature, to save some of the forests and soils—not for the oil and timber elite or even for the backpacking elite, not for social abstractions or for personal profit, but for our heirs, for them to see and decide to save or to use.

Action Responsibility & Wisdom
Now is the time to define goals in terms of population, quality of life, and preservation of biomes. Resolving conflict through social debate would free unprecedented resources to satisfy social needs. That which has been hitherto left unsaid—the goals of humanity-could become explicit. Goals are not some final state reached once and for all time, but a horizon. Eutopias offers continuity towards a horizon.

Solutions to uncertain futures can be found in the characteristics of good places and in the principles applied to them. The proper actions come out of common sense. We need to be sure that we allow ecosystems to regenerate healthy conditions while we take our needs from some of them. We need to plan for at least seven generations ahead, being flexible and keeping some options open, and being as self-reliant as possible. We need to be frugal with most resources and keep seven years of food and supplies in reserve. We need to identify an optimum population, over a minimum and maybe fifty percent below a maximum. We need to be as playful and joyful as any previous generation.

Science presents us with too many facts, yet we crave to have more. Philosophy presents us with too many values, but we attend to too few. Technology presents us with too many things, but we do not know what we need. We do not need more information or rules, but we do need meaningful ideas. Our attitudes and feelings toward nature need to be revitalized with evocative metaphors that let us accept responsibility for that part of the earth that we build, namely human culture and human landscapes. In order to know what is important and what is valuable, we need wisdom that we may not have.

The words 'view,' 'vision,' 'witness,' 'wise,' and 'idea' are all derived from the Indo-European word meaning to see, understand, or know. Wisdom is knowledge of the larger interactive system, which if disturbed, can generate exponential curves of change. Greed is unwise. Wisdom is recognition of and guidance by a knowledge of the total system. The system punishes any species unwise enough to quarrel with its ecology. Size and pride are unwise. Any course of action, like that just discussed, that ignores ecological stability and intentionality, i.e., the logic of nature, is unwise.

Hans Vaihinger (1911) in *The Philosophy of As If* suggested imitation as a solution to our lack of wisdom. Humans have no choice but to live by fictions; as if this world is the ultimate reality, as if there were free will. Humanity must plan for its future as if its days were not counted (or at least for several billion years). Jonas Salk urges us to behave 'as if' we were wise, by using good sense. Wisdom is a new kind of fitness. To survive, we must accommodate ourselves to the new conditions of a radically different life. Survival in this sense is not a win or lose proposition, but a double win.

Wisdom is knowledge of the larger interactive system, which if disturbed, can generate exponential curves of change. Wisdom is the recognition of and guidance by a knowledge of the total system. Lacking knowledge, lacking wisdom, we must behave 'as if' we were wise, as if we had good sense. Humans have no choice but to live by fictions, as if this world is the ultimate reality, as if we are responsible for our actions. Humanity must plan for its future as if its days were not counted (or at least for several thousand years). Wisdom is a new kind of fitness. To survive, we must accommodate ourselves to the conditions of the earth.

Wisdom is the disciplined use of the imagination with respect to alternatives, exercised at the right time and in the right measure. But we need practical wisdom, prudence, and intellectual control in virtue, in place of the theoretical wisdom taught by schools. The truths of our unique cultures and the wild earth are apprehended through myths. The poetic language of mythology can fit all the facts and values, things and images, into our hearts so that we can feel them and act upon them — so that we can make good places.

Related to wisdom, there are many corrective factors of human action: Contact, art, love, and religion. Love is the formation of Martin Buber's I-Thou relationships between human and society and environment. Socrates stated that Eros is midway between wisdom and ignorance. He who has no sense of his own deficiency will have no love of wisdom. Love is the desire that good be one's own for as long as it can. Arts and the activities of the mind can correct the excesses of pride. Contact between man and nature and animals can correct the problems of abstraction. And, ecodeontics (or religion), the binding of humans to the invisible powers of place, can correct the effects of detachment.

Starting Now

Will Eutopias work? Yes, if — and that word makes a difference. We can try to be wise, and act 'as if' it might work. Will we survive? Has some limit been exceeded? If not, do we have time to correct our actions? Time has become a real problem, especially since calendars were invented to keep track of big events. If the climate or our lives were more regular, we might not be so concerned with time. The idea of a regular or eternal return might be satisfying enough. If a fundamental limit has been exceeded, what should we do? If it is too late, and it is very difficult to know this definitely or absolutely, then we could grow until we crash.

A Eutopian framework addresses the inadequacies of the present system; it offers a drastic system change from the institutional gridlock of elitism, but the change is not so drastic that the feasibility of acceptance is too low. The benefits must be worthwhile to justify the costs. The benefits cannot be vague and unsatisfying when the costs are immediate and painful. Communication and education must prove that the benefits exist, so that the eutopian alternative can be exercised. It must be a participatory movement, and it must appeal to a large segment of the total human society. Since not all interests will be satisfied, there must be opportunities for transformations or for alternate paths.

The Eutopian framework is an open, flexible, and partially-planned global relation, instead of a finished, closed, completely-planned society, as imagined in utopias. A Eutopian framework accepts the imperfect nature of humans and the changing ambiguity of nature. Eutopias can detoxify cultural rivalries. Racism, sexism, ageism, and speciesism lose their importance in a cooperative society of advanced communication, automation, equality, humane scale, and meaningful preservation. We could act 'as if' we were wise. We should start now. It is an *emergency!*

Ecodex (Summary of a Eutopian Code)
As a framework for global emergency actions, Eutopias is a total reconsideration of the current *pattern* of technologies, cultures, values, systems, and behaviors, evolving into a low-profile technological ethic suitable for a renaissance. It is a code for preserving those parts of the earth that are needed for renewing the holecosystem and for habitats for the billions of animals, plants and living beings that are part of the earth. It is a code for allowing fair use of that part of the earth that is human. It is a code for human equality in opportunity and worth. It is the demand for a margin from catastrophe, so that if humanity is unable to live peaceably, the rest of the earth will not become extinct as well.

The new theme for people's minds may begin in prose, but it should culminate in poetry. The human mind, under pressure from the dialectic process, grows into more subtle noetic experiences, until ecstatic insight blossoms. We must learn to be an individual in a human society in an ambihuman ecology with amphibian grace. To paraphrase a line from Keats, the poetry of the earth is never dead, but is could become unreadable to us, and as remote as the stars.

What mysteries of the universe we cannot understand, we can accept in faith. The working out of the cosmic process is effected by the actions of human beings, by hate or love. Love is reverence for the experience of all beings. Through love, as well as effort and intelligence, humans can make good places on earth.

The interactions of billions of small actions cause a change in quality, that is, quality emerges from quantitative action. Thus, rare events can shape the entire course and texture of the planet and its systems. The unconnected actions of 7 billion people can really alter a change in scale that can result in environmental disasters. Of course there are other game changers also, including poverty and conflict. Every action is internal to the system. Human actions can create new risks and risk spirals to occur. Therefore, the more conscious and coordinated actions can be, the more effective they will be in responding to emergencies.

Many people are warning of imminent doom, and they are surely correct. Others are praising a glowing future, and they are certainly right. But, either way, we have to create those circumstances. We need use the information have and have the wisdom to apply it to our actions. We need to be determined to start acting. There are so many streams of information, so many different forces. There will be too much information and force, and too many options. We will have many choices, some easy, others less easy. Civilization has become complex and global, so its continuity requires us to work together on global problems and even for help on local problems. More people will be happier when we can work on a common future if they participate and shape it.

We need to place hot triggers everywhere to inspire *immediate* action! Triggers such as lowering a thermostat or planting trees. We need to promote sticky ideas and actions, with the characteristics of concreteness (we will *stabilize* the planet in 5 years), simplicity (we *can* do it by working together on a global plan), credibility (we *have already* created global associations and affected climate and environments), surprise (it can be a fun challenge and bring us together), and emotion (succeeding will make us confident and strong), within a gripping story (imagine telling how this grand, coordinated effort worked to your children). This will increase recognition and the likelihood of sharing. We need to nudge people into good behavior — by a default choice as a path of least resistance — without limiting incentives or restricting their options. We need to make sure the goals are easily seen and that personal incentives are attractive. People need to give and receive feedback as they participate.

We need to consider the negative aspects. Hunger, poverty, degradation, instability, disease, and war persist despite technical brilliance and personal empathy. If the system is skewed, dutiful actions can produce horrible results. So, we need to restructure the system or change systems. And, sometimes, if actions are taken too fast, they may amplify short-term variation and create unnecessary instabilities. From a wider perspective, information flows, goals, incentives and disincentives can be restructured so that bounded rational actions can add up to good results.

Social changes can occur very rapidly, however, when the time is right for them; for instance, oil-producing nations became the financial equals of industrial countries within months. An immediate, realistic, coordinated program of action is needed, capable of being implemented by communities and global agencies. We must face our responsibilities directly, declaring that there is no place in a eutopian society for monopolistic and multinational corporations, for the maniacal religion of merchandise, for genocidal military establishments, for urban explosion, for state socialism, for overbearing bureaucracies, or for technocratic politics; we must act to end them, now. The declaration must be political, through cooperative networks or leaderless consensus, by persuasion and example. The problem of human existence on the planet must be approached without deference to artificial boundaries of states, races, or castes. Poverty, pollution, repression, are concerns of every human community. We must stand and state that nature has limits, that we cannot have everything we desire.

What kind of actions are involved in dealing with a global emergency? What do they look like taken together? (See Table 1).

Table 1. Sample Binary Challenges & Responses

Complex Problem   Simple Solution

Addictions (cheap oil/credit/sugar/drugs): Discipline. Reset costs to reflect rarity and value. Require more for credit. Reward changes to health.

Agricultural failure: Stress permanent agriculture, gardens (e.g., Permaculture or Fukuoko). Shift subsidies from milk and corn to low-impact, organic. Decentralize operations to small personal farms.

Anger at system: Improve without destruction. Train in meditation or support groups.

Arrogance: Realistic self-assessment. Education of surrounding complexities. Slow down. Accept change and limits.

Architectural failure: Stress green design, small homes, impact fees, true costs. Improved LEED standards. Make buildings wildlife friendly. Combine functions to deemphasize iconography.

Bad designs: Weed out dangerous, unrelated designs. Stress ecological design process.

Biodiversity loss: Conservation. Wilderness restoration. Monitor species and habitats.

Catastrophes (collapses/accidents): Global/national savings accounts. Monitoring. Replacement of endangered settlements. Set stage for renewals.

Cheating (financial, social, economic): Tax currency transactions. Ban corporate contributions to elections. Fix progressive taxes. Constant scrutiny. Strictly regulate economic and financial institutions.

Climate chaos: Set global goals. Start climate trust. Reduce emissions. Reduce carbon with tax incentives. Accept change gracefully; respond to it.

Community collapse: Strengthen communities; keep money local. Tax sprawl and land, not buildings or income.

Conflicts/violence: Engage in conversation. Adhere to cultural or international rules for disputes. Use third-party help.

Corporate misbehavior: Require better constitutions. Remove personhood. Set size/location limits.

Deforestation: Value wood, trees & forests. Replant native species. Create goals plans.

Dependence, unhealthy: Stress self-reliance and individual abilities. Encourage independence and partnerships.

Discrimination: Cultural representation. Educate for understanding.

Drawdown resources: Impose limits on amounts and rates of resource use.

Ecological Collapse: Impose standards and monitoring. Limit system exploitation. Prepare for restoration or renewal.

Ecological destruction: Apply loss taxes. Oliver Wendell Holmes noted that taxes were the price of civilization & investments in the future. Monitor ecosystem 'services.'

Energy waste: Reduce fossil fuel dependence. Diversify alternative sources. Limit and tax use. Promote efficiency. Reuse low-grade (in industrial ecosystem). Decentralize.

Ennui / apathy: Stress importance of participation. Volunteerism. Restart Civilian Conservation Corps. Education in values and diversity. Awareness of catastrophe. Inculcate responsibilities.

Extinction/endangerment: Big Deep Wilderness. Conserve, secure habitats. Triage for endangered species. Stop destroying established natural patterns of renewal.

Financial cheating (Failure): Everyone pays dues. National & global savings accounts (from taxes & fees) to limit failures.

Fragmentation: Apply comprehensive, holistic thinking to human infrastructure. Reduce conversion, expansion and road-building.

Frustration/panic: Accept uncertainty, error and regular waste. Plan for it. Restore confidence in society and government.

Globalizing chaos: Represent nations of cultures. Create global government. Fund it from commons taxes on global resources. Create usable, accessible information structure. Make plans and designs.

Greed/lust: Ban inappropriate advertising. Limit bank size/purpose. Rules.

Growth runaway / Mindless progress: Stop growing. Stress development. Analyze and assess real progress with benchmarks.

Ignorance: Make education fun and valuable. Support every level. Encourage imagination and consciousness of the whole, as well as actions.

Inequity: Redistribution ownership. Create form for equitable exchanges based on human labor. Use cooperation model. Reconstruct economics with more reciprocity. Display outrage at historical patterns.

Illness (subhealth): Development and play. Physical education, universal health-care. Tax incentives for healthy behavior. Healthier attitudes in healthier societies and environments. Improve sanitation infrastructure.

Isolation: Build strong communities of residents and interests. Voluntary exchange. Create partnerships with groups. Create open patterns.

Luxuries (unsustainable): Impose limits and taxes. Educate on realistic goals. Allow social luxuries instead. Limit dangerous ones, like cars.

Monopolies: Impose corporate rules and limits. Break down.

Overconsumption: Discipline. Tax excess. Educate on limits. Ration some things.

Overexploitation (Overshoot/Overfishing): Impose use taxes, strict limits. Establish marine reserves, ecomanagement, scientific fish catch regulations.

Political Domination: Encourage/reward participation. Equalization of opportunities. Require tests. Limit service to single 3-yr terms. Install large global government.

Pollution/toxins: Adjustment Taxes on Bads. Phase-out. Industrial ecology. Conserve resources. New infrastructures. Taxes on pollutants. Limit synthetics, solve plastics degradation. Recycle heavy metals.

Population growth: Plan population based on carrying capacities. Issue license vouchers. Educate on family. Increase equities.

Poverty: Level playing field. Offer economic assistance. Promote equity.

Refugees: Environmental restoration of places and homes. Location and relocation assistance. Peace process.

Shortages (water/wild/food/energy): Value water. Restore watersheds, flood protection. Recycle regularly. Limit waste in agriculture and industry. Ration necessities according to cultural carrying capacity

Technology, runaway: Use precautionary principles. Set goals and limits. Divert momentum for growth and excess. Create goals and limits for technology, e.g., precautionary principle. Integrate technology with design links and rules.

Threats (ecological/climate/extrasolar): Monitor planet and space. Pay attention to the distant future (and past).

Transportation waste: Fund public transport, rail, sailing ships, buses, bicycles. Reduce necessity for compulsive movement.

Under/unemployment: Apply ecological economics. Buy and sell local. All jobs part-time. Pay parity. Living wage. Work-sharing. Minimum vacation.

Small business tax credits. Address skills deficits with education. Strengthen labor unions in relation to trade and management.

Uniformity/sameness: Encourage diversity. Emphasize personal and cultural identities and customs.

War/aggression: Disarmament of nuclear and large-scale weapons. Peace-keeping forces. Cooperative peace after UN disarmament. Equity. Partnerships. Trade specialization.

These responses to challenges are individually simple and based on common sense. Job creation, for instance, is critical for full employment, but it is equally critical for an emergency transition to balanced planet; there is so much to be restored and rebuilt. Other responses, such as arcologies or clipper ships, are fun and exciting. Some, like peace work, are overdue. Others, like rewilding the ocean, will require heroic sacrifice. All of them will benefit future generations and the planet. To minimize untested efforts, these actions are based on the values and forms of traditional cultures. Due to that and modern communications, actions could be rapid (See Table 2 for time estimates). Rational planning could catch up as it developed.

Table 2. Several Time Estimates

Engage all people	33 days	*Empower UN or GU*	26 days
Invite New Nations	4 days	*Complete disarmament*	7 days
Claim Global Commons	3 days	*Create new laws*	9 days
Set Goals & Standards	6 days	*Produce plans & designs*	25 days
Set up UN Departments	7 days	*Make budgets*	8 days
Coordinate nations	23 days	*Shift agriculture to eco-*	11 days
Address equity issues	18 days	*Start restorations/wilds*	5 days
Start jobs programs	15 days	*Start education programs*	15 days
Start new cities	17 days	*Start reforming economics*	12 days
Total Soonest	**173 days**		

Think these estimates are unreasonably short? Events can happen in days; witness the collapse of Communism as a form of government in Europe (1989). Ideas can go viral in days; consider the Arab Spring coordination by social media computer networks (2012). Creating departments, budgets and industrial reassignments in days has happened previously; remember the US response to the attack on Pearl Harbor (1941). Large numbers of advisors can be moved quickly to areas of catastrophes; witness the response to Ebola outbreaks (2014). Furthermore, many of these goals, standards, plans, and designs already exist on paper and in the minds of many, and need only to be debated and refined (or replaced) by other people participating in a coordinated global effort. Keep in mind, many of these activities overlap (hence the smaller total shown—173 days).

Obviously, the number of days is a fuzzy number that would become set by events, but these calculations are based on many other historical events and on the calculations of others, e.g. Earl Osborn and Leopold Kohr. Rather than debating the correctness of estimates, energy put into bringing the actions into play could beat the estimates, or could take as much time as was needed. Remember that tens of millions of people are suffering and dying in a matter of days, due to problems with distribution and government, so it is urgent to act.

Central to this effort to outline and initiate a participatory frame is the recognition that our catastrophic situation requires *immediate emergency actions* on local, regional and global scales, from expanding personal consciousness to reforming the very character of human civilization and its coevolution with wild nature. An international organization, such as the United Nations or a new Global Union of Commonwealths, has to assume control of the planetary commons. It would manage the commons in the interests of all living communities. By charging usage, extraction or loss fees for water, fossil carbon and other resources, it would become self-supporting. By international agreement, it would have the largest and only standing army, with some large-scale nonnuclear weapons (nations and communities would maintain police forces for public safety and control). It would allow all nations—even minor landless ones or some unique business entities like corporations—to join with equal votes. It would protect traditional and modern cultures in a loose framework, encouraging cultural health related to environmental health, stressing economic equity through a variety of measures, and trying to direct the emerging global culture. This can be done immediately if we try, and so much would be gained. And, if we do not try—or if we do and fail—what more could be lost?

Figure 8. Plan of Several Northern Hemisphere Wolf Paths

Appendix

This section contains specific notes from the text, as well as definitions of terms, a bibliography, an index of subjects and names, and author notes and biography.

End Notes

[1] Some of these numbers are from Rudolph J. Rummel and are estimates.
[2] From Michael W. Fox, HSUS, 1984.
[3] From J. Cousteau et al., 1984.
[4] From N. Myers et al., 1984.
[5] Borgstrom, G. 1975.
[6] Osborn, E. 1962.
[7] Klein, D. 1970.
[8] "The state of human development," United Nations Development Report 1998, Chapter 1, p. 37.
[9] T. H. Marshall, 1965.
[10] F. LaMarche. 1973.
[11] In his 1964 book *Architecture Without Architects*, a short introduction to nonpedigreed architecture, based on his MoMA exhibition.
[12] Fed Reg V70 N33 Pps 8373-5.
[13] Edward Wilson p. 185.
[14] in *Defining Sustainable Forestry*
[15] Milton Friedman et al.
[16] Franklin, J. 1988.
[17] Kohr, L. p. 196.
[18] With help from Arne Naess.

Glossary, Bibliography & Index

Found at the website: www.Eutopias.net & SynGeo.org

Author's Note

This book was taken from a much larger, unfinished work-in-progress, *Eutopias*, which is available as a draft on its website, www.eutopias.net. The sections in that work are numbered sequentially, but the numbers are not kept for this book. There are, alas, a few misspellings, some bad grammar and many unfinished thoughts. Some of the ideas in this work were radical forty five years ago, but have thankfully become acceptable or commonplace. Others are still considered awkward or unpalatable to professional journals. I ask you to participate in and improve this conversation despite these flaws and shortcomings. Thank you for your consideration.

To make up for the loss of trees and their services, as a result of my use of paper in this books, I have planted over ten thousand trees, during a period of twenty years, at the Altazor Forest in Idaho. More plantings are planned in Oregon, Idaho and Virginia forests.

Biography
Alan Wittbecker is Senior Design Ecologist with SynGeo ArchiGraph LLC in Florida, where he works on a series of projects in global ecological design. He also teaches Forest Science at the Ecoforestry Institute and Environmental Science at Ringling College. A veteran of the U.S. Air Force, Wittbecker is also a returned Peace Corps Volunteer from Bulgaria, where he monitored wolves in the Central Balkan Mountains. He has used his education and interests to explore a spectrum of ecological applications, from research on forest 'pests' — larch casebearers, cedar powderworms, coyotes, and bears — to the political implications of the protection of species and habitats. When not engaged in preservation activities or thought experiments, he enjoys walking, swimming, reading, and drawing, at the River Farm Forest. You can reach him at: home@eutopias.net

Colophon

Type: Book Antiqua 10pt
Display Type: Book Antiqua 12pt
Book Design: Rian Garcia Calusa Designs
Cover Design: Rian Garcia Calusa
Graphics: Alan Wittbecker
Author Drawing: Merissa DePasse
Editing: J. Garcia B. of Rian Garcia Calusa
Hardware: Macintosh G5, HP 3310
Software: Photoshop, AzTex, InDesign & Acrobat
Furious Charge & Entertainment: Ving Ringneck Snake
Spiritual & Material Support: Precious Woulfe

www.ingramcontent.com/pod-product-compliance
Lightning Source LLC
Chambersburg PA
CBHW051649170526
45167CB00001B/392